WOLVES, BEARS, AND THEIR PREY IN ALASKA

Biological and Social Challenges in Wildlife Management

Committee on Management of Wolf and Bear Populations in Alaska

Board on Biology
Commission on Life Sciences

National Research Council

NATIONAL ACADEMY PRESS
Washington, DC 1997

NATIONAL ACADEMY PRESS • **2101 Constitution Ave., NW** • **Washington, DC 20418**

NOTICE: The project that is the subject of this report was approved by the Governing Board of the National Research Council, whose members are drawn from the councils of the National Academy of Sciences, the National Academy of Engineering, and the Institute of Medicine. The members of the committee responsible for the report were chosen for their special competencies and with regard for appropriate balance. In preparing its report, the committee invited people with different perspectives to present their views. Such invitation does not imply endorsement of those views.

This report has been reviewed by a group other than the authors according to procedures approved by a Report Review Committee consisting of members of the National Academy of Sciences, the National Academy of Engineering, and the Institute of Medicine.

This study by the National Research Council's Commission on Life Sciences was sponsored by the National Institutes of Health and the National Science Foundation under contract no. N01-OD-4-2139. Points of view in this document are those of the authors and do not necessarily represent the official position of the sponsoring agencies.

Library of Congress Catalog Card Number 97-75390
International Standard Book Number 0-309-06405-8

Additional copies of this report are available from:
National Academy Press
2101 Constitution Ave., NW
Box 285
Washington, DC 20055
800-624-6242
202-334-3313 (in the Washington Metropolitan Area)
http://www.nap.edu

Cover art created by Vivianna Padilla.

iii

The National Academy of Sciences is a private, nonprofit, self-perpetuating society of distinguished scholars engaged in scientific and engineering research, dedicated to the furtherance of science and technology and to their use for the general welfare. Upon the authority of the charter granted to it by the Congress in 1863, the Academy has a mandate that requires it to advise the federal government on scientific and technical matters. Dr. Bruce Alberts is president of the National Academy of Sciences.

The National Academy of Engineering was established in 1964, under the charter of the National Academy of Sciences, as a parallel organization of outstanding engineers. It is autonomous in its administration and in the selection of its members, sharing with the National Academy of Sciences the responsibility for advising the federal government. The National Academy of Engineering also sponsors engineering programs aimed at meeting national needs, encourages education and research, and recognizes the superior achievements of engineers. Dr. William Wulf is president of the National Academy of Engineering.

The Institute of Medicine was established in 1970 by the National Academy of Sciences to secure the services of eminent members of appropriate professions in the examination of policy matters pertaining to the health of the public. The Institute acts under the responsibility given to the National Academy of Sciences by its congressional charter to be an adviser to the federal government and, upon its own initiative, to identify issues of medical care, research, and education. Dr. Kenneth I. Shine is president of the Institute of Medicine.

The National Research Council was organized by the National Academy of Sciences in 1916 to associate the broad community of science and technology with the Academy's purposes of furthering knowledge and advising the federal government. Functioning in accordance with general policies determined by the Academy, the Council has become the principal operating agency of both the National Academy of Sciences and the National Academy of Engineering in providing services to the government, the public, and the scientific and engineering communities. The Council is administered jointly by both Academies and the Institute of Medicine. Dr. Bruce Alberts and Dr. William A. Wulf are chairman and vice chairman, respectively, of the National Research Council.

Preface

Management of wildlife in Alaska is carried out by the Alaska Department of Fish and Game (ADFG), Division of Wildlife Conservation, guided by policies established by the Board of Game (BOG), which has constitutional and statutory authority to set game policy direction and to regulate game harvest and management. Most management actions implemented by ADFG are not controversial, but wolf control and management in Alaska have become increasingly difficult because public perceptions of the roles and values of predators and the ethics of predator control are changing. Conflicts among people with different interests in wolves are intense.

The governor of Alaska, Tony Knowles, suspended the state's wolf control program in late 1994 because he judged it to be unacceptable in its treatment of wolves, as well as nontargeted species. He stated that he would not reinstate predator control unless it met these tests: (1) it must be based on solid science; (2) a full cost-benefit analysis must show that it makes economic sense for Alaskans; and (3) it must have broad public support. In July 1995, Governor Knowles asked the National Academy of Sciences (NAS) to undertake a scientific and economic review of management of wolves and bears in Alaska. In July 1996, the National Research Council, the operating agency of the NAS, established the committee on the Management of Wolf and Bear Populations in Alaska.

The committee's mandate was to synthesize what is known about the biological aspects of wolf and bear management in Alaska, with particular emphasis on the degree of certainty one can have about predictions about the impact of wolf or bear management on both predator and prey populations. In addition, the committee was asked to identify additional biological data that should be collected when planning, conducting, or evaluating predator management programs. (We use the terms management and control as defined by ADFG to include alterations in populations by any method, whether by killing, translocation, or diversionary feeding.)

As part of the economic review, the committee was asked to evaluate the methods that are most relevant for assessing the costs and benefits of predator management programs in Alaska, and to identify what types of data should be collected for an appropriate economic analysis of predator management.

The committee used a variety of methods to assemble the information it needed to carry out its mandate. It held four meetings between September 1996 and March 1997, including forums in Anchorage, Fairbanks, and three villages in the Alaskan interior. Input was solicited from a wide variety of organizations representing hunters, Alaskan Natives, trappers, animal welfare advocates, and tourism associations. The committee assembled, analyzed, and interpreted existing scientific literature on dynamic relationships among wolves, bears, and their prey, and control and management programs in Alaska and other high latitude regions. The committee analyzed all available information on the economic values of consumptive and nonconsumptive use of the focal species, together with data on public attitudes towards predators, their prey, and the goals and methods employed in predator management and control efforts.

The committee sought to show how to improve the ability to predict the outcome of implementing a management program, and to increase public confidence that decisions were being made using all available relevant information, and that the information was analyzed and interpreted appropriately. In addition, the committee assessed which information collected during a management program would be most valuable in removing uncertainties that have important policy implications. The committee emphasizes, however, that it makes no recommendations about whether predator control should or should not be carried out. To attempt to do so would go well beyond its mandate. Whether and when to control or not control predators is a policy decision to be made by the BOG, based on input from the public and recommendations and data provided by State and Federal and agency personnel. The role of the committee was to advise on the ways that scientific, socioeconomic, and decision-making data can best be used to assist managers to make wise decisions.

To carry out its mandate, the committee gathered information on the goals and objectives of wolf control and wolf and bear management in Alaska, the extent and nature of local knowledge of predators and prey, and the population dynamics of mammalian predators and their prey in managed and unmanaged northern ecosystems. The committee analyzed past predator control and management programs in Alaska and other northern regions to determine the biological consequences of actions carried out under different conditions and using varied methods. The committee also evaluated studies of the attitudes of all segments of the Alaskan population toward predator control and management and the methods used to accomplish it, and analyzed what was known about the economic value of consumptive and nonconsumptive use of wolves, bears, moose, and caribou.

In its work, the committee was assisted by many people, among whom are those who addressed the committee at its public sessions in Anchorage and Fairbanks and wrote letters in which they shared their views and information with us. Our understanding of the different needs and perceptions of native Alaskans was improved by meeting with people in Aniak, McGrath, and Galena. We thank them and the people who helped arrange our visits to those villages. ADFG biologists, social scientists, and managers were particularly helpful in providing access to their data, some of which were unpublished, and they responded quickly to help us verify facts and track down published papers and unpublished manuscripts. Committee members also met and corresponded with scientists from the US Fish and Wildlife Service, the National Park Service, the Yukon Territorial Government, British Columbia Ministry of Environment, and independent scientists to become familiar with the full array of information that has been gathered by the many scientists, social scientists, and managers who have worked on wildlife management problems in Alaska and northern Canada.

This report was improved by a group of anonymous reviewers who provided the committee with prompt and insightful evaluations of the semifinal draft of the report. All members of the committee gave generously of their time to assist writing the report and to engage in the extensive discussions that were necessary to achieve consensus on a wide range of complex issues. In our work, we were ably assisted by NRC staff Janet Joy, Jeff Peck, and Allison Sondak. Their care and tending of the committee contributed much to our efforts and our final report. Janet in particular worked overtime to bring this report to a timely completion while she changed positions at the NRC.

Gordon H. Orians, *Chair*

Contents

TABLES

FIGURES

WOLVES, BEARS, AND THEIR PREY IN ALASKA

Executive Summary

The primary goal of predator management in Alaska is to increase prey populations for human harvest. Governor Tony Knowles suspended Alaska's wolf control program in late 1994 and stated 3 criteria for the reinstatement of wolf control: it must be based on solid science; it must make economic sense for Alaskans; and it must have broad public support. The Committee on Management of Wolf and Bear Populations in Alaska was asked to evaluate the biological underpinnings of predator control and management in Alaska and assess its economic impacts. Such an evaluation is necessary, but nonetheless not sufficient to satisfy Governor Knowles' stated criteria. In particular, the committee was not asked to assess whether wolf control should be conducted in Alaska nor was the committee asked to design a program that would have broad public support. Both issues are political matters that will be decided by the Alaskan public and their elected officials.

The evaluation of the biology and economics of predator management requires understanding of the ecological, economic and political contexts in which the program is carried out because all these components determine which programs are feasible and effective. One must have some measure of how much more prey people feel they need and if they are willing to devote the resources needed to achieve the goal. How long should predator control be in effect, by what methods, over how large an area, and how many animals need to be removed? Finally, any predator management program must be planned in such a way that its impact on the adult prey population can be measured. Thus the report first describes the socioeconomic and ecological context for wildlife management in Alaska. This is followed by a discussion of the biological aspects of wolf

1

and bear management, a discussion of the kinds of economic analysis that is appropriate to assessing the impact of predator management programs, and an analysis of decision-making processes. The final chapter presents the committee's conclusions and recommendations.

PREDATOR CONTROL AND MANAGEMENT: PAST AND PRESENT

Predator control in Alaska has paralleled that elsewhere in the United States. During the first half of the century, predator control was seen as important for both human safety and protection of prey, especially livestock. The goal of most research on predators was to determine how to reduce their numbers. Data on their life histories, their ecological roles, and predator-prey relationships were rarely collected. As public attitudes shifted during the 1960s and 1970s, predator control programs were increasingly questioned and more broadly-based data were gathered on predator ecology. In the 1980s and 1990s, the public increasingly demanded active management of nongame species. Controversies over predator control and management continue today in many states.

Wildlife management policies in Alaska are set by the Board of Game (BOG) whose members are appointed by the Governor and require confirmation by the state legislature. Management of wildlife in Alaska has been carried out by the Alaska Department of Fish and Game's (ADFG) Division of Wildlife Conservation. The BOG holds two formal meetings each year in each region, during which all the regulations for that region are considered and public input is solicited. Most management actions implemented by ADFG are not controversial, but wolf control and management have become increasingly difficult because public attitudes toward predator control are changing.

ALASKA'S PEOPLE, BIOMES, AND WILDLIFE SPECIES OF CONCERN

About 610,00 people live in Alaska, most of them in and around Anchorage, Fairbanks, Juneau, and a number of smaller south-coastal cities. Alaskan Natives comprise 16.5% of the state's population and most Alaskan Natives live in rural communities. About 80% of the non-indigenous residents of Alaska live in urban communities. About 60 percent of the land is in federal ownership, 28% is in state ownership, and 12% are Native lands. Only 1% is in private, non-Native land. Human impacts on Alaskan ecosystems are much less than in other parts of the United States. In northern Canada and Alaska, habitats and the large mammal system are largely intact. Alaska's biomes are diverse, ranging from temperate rainforest to arctic tundra. Wildlife management activities are correspondingly diverse but most predator control efforts have been carried out in the interior boreal forest region.

Wolves, bears, and their primary prey differ strikingly in their characteristics and requirements and their responses to management. In Alaska, wolves are found over most of their historic range at densities that are strongly correlated with variations in ungulate biomass. Wolves disperse over great distances and readily colonize new habitats as prey become available. In most packs, only one female reproduces each year. Consequently, populations with larger packs have a lower proportion of breeders than populations with smaller packs.

Wolves die of a variety of natural factors, such as starvation, accidents, disease, and intraspecific strife, and these factors may cause annual mortality rates as high as 57% in some years. Under favorable conditions, wolf populations can increase as rapidly as 50% per year, and rates of increase are often higher after heavy exploitation because per capita food availability is typically high under those conditions. Because of their high reproductive rates, wolves can usually maintain their populations even if annual mortality rates are as high as 30% of the early winter population. For the same reason, wolf populations typically rebound to pre-control densities within 4-5 years after termination of control efforts.

Black bears are closely associated with forests, but brown, or grizzly, bears are larger, live in more open country, and are usually found at much lower densities than black bears. Although the distribution of both species has dramatically decreased since European settlement of North America, they survive in most of their historical range in Alaska. During the part of the year when they are active, bears spend most of their time feeding. Bears are relatively efficient at catching the newborn ungulates in spring. Bears have long interbirth intervals and are slow to mature. The low reproductive rate of bears is balanced by high annual adult survival rates, but because of their low reproductive rates, bear populations are easily reduced by hunting and recover slowly. Brown bear densities in Alaska are highest in coastal regions where abundant salmon and good vegetation are available for them to eat. Male bears spend most of their lives as solitary individuals, whereas females are either alone or with their latest litter of cubs. Because bears move over large areas, and retire to their dens in winter when they would otherwise be easy to count, they are difficult to census and estimates of their densities are poorer than for wolves.

Caribou inhabit the high Arctic, tundra, boreal and sub-boreal forests, and wet interior mountains across much of the Northern Hemisphere, and are found in most regions of Alaska. Most caribou live in herds that annually move over extensive areas, sometimes migrating hundreds of kilometers between wintering areas and calving-summering grounds. Female caribou usually show strong fidelity to the specific calving grounds of the herd, returning each year to the same general area to give birth. Caribou are primarily grazers. They eat a variety of plants during summer, but subsist on lichens during winter.

In contrast, moose are solitary, nonmigratory browsers that eat primarily leaves and tender stems of woody plants. An area that has burned 25 years earlier provides ideal moose habitat. Many female moose give birth to their first calf at two years of age and twinning is common during subsequent pregnancies. Caribou do not reach breeding age until their third year and usually give birth to a single calf per year. Therefore, moose populations can recover from depression faster than caribou if the quality and quantity of forage is adequate. The high survival rates of adult female moose and caribou, which are usually above 85% per year, mean that these species can thrive even when calf mortality is substantial.

PREDATOR-PREY INTERACTIONS

It might seem obvious that, if predator numbers are reduced, prey numbers should increase, but predator-prey interactions are not that simple. Predators and prey interact in an open system, subject to many influences. Prey numbers are influenced by weather and habitat quality, primarily food quality and abundance. Not surprisingly, models of predator-prey interactions predict a wide array of results depending on characteristics of predators, prey, and the environment in which they interact. The major results of predator-prey models most relevant to wolves and bears and their prey are the following:

• The removal of predators from a plant-herbivore-predator interaction system can either stabilize or destabilize herbivore population dynamics.
• Two alternative stable states may exist in predator-prey systems, with a lower equilibrium corresponding to very low prey and predator populations, and a higher equilibrium corresponding to high predator and prey populations (with the prey close to their carrying capacity).
• If a regression analysis is used to determine what controls prey populations in a predator-prey system, the factor that explains the greatest proportion of the variance in prey population growth rates depends largely on where "noise" enters the system, and not on what actually controls the dynamics.
• Correlative studies have limited value in inferring causation.
• The interactions between prey and their plant resources need to be understood.
• The task of identifying which "model" describes a particular situation is technically challenging.

Because of these challenges, analysis of predator control in Alaska is a major scientific task, and long-term management could be improved if a more solid understanding of wolf-caribou or wolf-moose interactions were available. Nevertheless, some of the long-term data sets on wildlife populations in Alaska are among the best available anywhere. ADFG scientists regularly publish their data in peer-reviewed literature, so their data and methods of analysis are carefully scrutinized by scientists elsewhere.

WOLF AND BEAR MANAGEMENT:
EXPERIMENTS AND EVALUATIONS

In addition to many studies on the population ecology of wolves, bears, moose, and caribou, a number of experiments have been conducted in Alaska and elsewhere in which wolf and/or bear numbers were reduced or otherwise altered and responses of caribou and/or moose populations were monitored. These experiments provide the best data with which to evaluate the biological basis of predator control as a management tool to increase ungulate numbers. In most of these experiments, wolves were killed, but some used translocation or diversionary feeding of wolves, bears, or both. Each of these predator reduction experiments was initiated because it was believed that decreasing the number of predators would result in increased ungulate densities and, as a consequence, increase hunter success. There was also the hope that higher densities of both predators and prey might persist for many years after predator control stopped. Each experiment began when ungulate densities were low or declining. Hunting was sometimes simultaneously reduced.

Although considerable time and effort was expended to plan, design, and implement these experiments, the results are less informative than might have been hoped. Part of the problem is due to the size of the areas in which the experiments were carried out, the difficulty of gathering the needed information, and budgetary limits. Further, some of these experiments were primarily management actions that were based on particular assumptions about predator-prey dynamics but were not designed to test those assumptions. As a result, less has been learned from these experiments than would have been possible had they been better designed and executed, and if the results had been more extensively monitored.

Although considerable data were gathered on population densities and trends of moose, caribou, and wolves prior to reaching a decision to initiate a control action, data on bear population densities were generally poor. In addition, assessments of habitat quality were limited primarily to indirect indices of ungulate nutrition (body weight, fat deposits, bone growth, pregnancy rates, calf:cow ratios). As a result, in most cases, data to support the judgments that habitats in the control area could support increased moose and caribou populations for more than a few years were limited. Several control experiments failed to increase ungulate populations, possibly because predation rates by bears were high, habitat quality was poor, or the area and duration were insufficient.

The degree to which, and the duration over which, wolf numbers were reduced varied considerably among experiments. In addition, wolf control was sometimes accompanied by other changes, such as reduction or elimination of hunting and trapping. The presence of such confounding variables makes it impossible to determine the relative contributions of the factors that were altered to changes in ungulate population densities.

Surprisingly, given the expense of conducting control experiments, the results of most experiments have been poorly monitored. In only two cases (wolf control on Vancouver Island, BC and Finlayson, Yukon) was there an attempt to measure whether hunter success increased, and it did in both. Only five experiments included measures as to whether ungulate population densities actually increased following predator control actions, but in 4 of these ungulate numbers did increase. During and after the other control experiments only indirect results, such as calf:cow ratios, were assessed, and sometimes they were determined for only one or two years.

Political pressures have created conditions that favor attempts to achieve quick, short-term results from predator control experiments by altering more than one factor simultaneously. In addition, budgetary constraints have led to the use of indirect measures of success that are less expensive in the short-term but which are not good indicators of population trends. ADFG did not measure hunter behavior before and after wolf control and cannot empirically show changes in where people hunted, their success per unit effort, and their satisfaction about changes in game densities that may have resulted from predator control efforts.

There are presently insufficient data to assess whether non-lethal methods of predator management are effective. It is too early to evaluate the effectiveness of sterilization of wolves in terms of its impact on ungulate populations. Diversionary feeding to divert wolf and bear predation on ungulates, a time-consuming and expensive process, has yielded equivocal results.

SOCIAL AND ECONOMIC IMPLICATIONS OF
PREDATOR CONTROL

Economic impacts depend, to a large degree, on public values. In democratic societies, government programs should be based upon widely shared public goals that legitimize the actions taken and the allocation of public financial and human resources to those activities. Funds for wildlife management are largely from license sales and federal aid from excise taxes on firearms and ammunition. Thus many hunters believe that their interests should be a primary concern in wildlife management. Nevertheless, some general public funds are appropriated for wildlife management. Their expenditure implies that the general public also is willing to pay for wildlife management.

Analysis of the social and economic dimensions of wildlife management programs must recognize that people differ in their attitudes, their values, and their economic situation. Consequently, there will be differences in how they view those programs. Alaskan Natives, other Alaskan residents, and people outside Alaska who visit there or might otherwise take an interest in its wildlife will react to wildlife management programs in distinctive ways and their reactions will affect the Alaskan economy in correspondingly distinctive ways.

Current attitudes toward predator control are varied. Most people disapprove of indiscriminate population reductions, poisoning predators, denning, or aerial gunning. Most Alaskans approve of hunting and a majority purchase hunting licenses at some time during their lives. Wolves and bears in Alaska also have high nonconsumptive values and most Alaskans desire more state managed areas for wildlife viewing.

For two reasons, the cost of a wolf control effort is likely to increase substantially if prevailing attitudes among the Alaska general public, especially urban residents and nonhunters, persist. First, most of the public objects to predator control except in specific geographic areas where a significant decrease in ungulate populations has been clearly and convincingly related to wolf and/or bear predation, and where significant economic or cultural hardship has been inflicted on resident hunters. Demonstrating and communicating this level of impact will inevitably increase management costs. Second, most of the Alaska public objects to control methods perceived as inhumane or unfair. Acceptable control methods tend to be more costly, labor intensive, time-consuming, or technically challenging. At the same time, the economic impacts resulting from public objections to wolf control efforts might be partially offset in areas where there is widespread support for wolf control among local peoples who derive economic and practical benefits from the activity.

A majority of nonconsumptive wildlife users and urban wildlife enthusiasts believe that intensive wolf and bear control reflects a well-entrenched bias among Alaska game management officials and hunter interests. This view encourages the use of legislative and judiciary methods for promoting policy change, rather than trying to exercise influence on administrative agencies. The success of the 1997 ballot initiative involving aerial hunting of predators will likely reinforce this perception. These views are likely to result in substantially greater management-associated costs, because of the heavy economic burden created by using legislation and litigation as a means of achieving policy goals.

Hunting and tourism provide an important contribution to the Alaskan economy in terms of both employment and expenditures. There are two reasons why data on employment and expenditures can be somewhat misleading for the purpose of assessing the economic benefits of predator control and/or the economic consequences of a tourism disruption triggered by public opposition to wolf control. First, although lay discussions of economic effects focus on jobs gained or lost, the more relevant focus is the increase or decrease in employment that would be associated with implementing or canceling a predator control program. *Total* versus *marginal* economic impact is an important economic distinction. The total impact measures the entire contribution of a sector. The marginal impact measures the increment or decrement in contribution associated with some expansion or contraction in the level of activity within the sector. If what is at issue is the total elimination of a sector, whether hunting or tourism, the total impact is the appropriate measure to use. But, if the change is smaller than that, then the correct measure is the marginal impact of the change. For example, if it involves some increment in hunting, or some decrement in tourism, then the appropriate quantity is not the total value of all hunting or all tourism, but rather the value of the particular increment or decrement anticipated to occur. This is harder to determine because it involves forecasting the specific magnitude of the increment or decrement in activity, as well as assessing the economic value associated with this change.

Second, expenditure is generally not a correct measure of value and does not reflect the broader social perspective in the sense of resources used up or changes in people's well-being. There is a basic distinction between financial or accounting revenues or costs and economic benefits or costs. Accounting revenues and costs reflect an assessment of monetary inflows and outflows according to the principles and conventions of certain accounting rules. Economic costs and benefits are intended to reflect a broader social perspective based on real costs, in the sense of resources actually used up, and real benefits, in the sense of actual changes in people's well-being. Although not entirely unrelated to financial revenues and costs, economic benefits and costs represent a different standpoint for assessment.

If a predator control program actually increases moose or caribou populations, the balance sheet of the potential impacts would consist of three main items: the marginal benefit of whatever increment occurs in the populations of prey species such as moose and caribou; the marginal cost of planning and implementing the predator control program itself; and any marginal reduction in benefit associated with the decrement in the predator population. The benefits can be subdivided into five components: (1) The gain in utility (that is, satisfaction or enjoyment) for residents from increased recreation associated with moose and caribou. (2) The gain in utility for nonresident from increased recreation associated with moose and caribou. (3) Any other gain in utility for residents arising from the increment in moose and caribou populations. (4) Any other gain in utility for nonresident arising from the increment in moose and caribou populations. (5) The gain in personal income from employment and profits for residents resulting from increased recreation or tourism in Alaska by residents or nonresident.

The costs similarly can be subdivided into: (6) The costs to government agencies and others for planning and implementing the predator control program. (7) Any loss of utility for residents associated with the reduction in predator populations. (8) Any loss of utility for nonresident associated with the reduction in predator populations. (9) Any loss in personal income from employment and profits for residents resulting from reduced tourism in Alaska by nonresident triggered by their reaction to the predator control program.

There are two core problems in assessing the economic benefits for hunting and wildlife viewing that might result from any increase in ungulate populations: predicting the change in recreational behavior—both the total level of recreational activity and its allocation among sites—and estimating the increment in consumer's surplus—the difference between what an item is worth to a consumer and what he actually pays for it—from recreation that is associated with this change in activity. Neither of these can be assessed from biological models of animal populations. Instead, both require a model of people's preferences and behavior.

In addition to the impact on recreation involving predator and prey species, and on personal income from employment and profit in Alaska, people may be affected by predator control programs in various other ways that reflect their concerns and involvement with these animals. Both residents and nonresidents might experience a gain or loss of utility from a predator control program for reasons separate from their interest in recreation, meat, hides, or other such uses of the animals. For example, they might care for the prey species and want it to be preserved regardless of whether they themselves plan to view it, hunt it, or eat it. Or they may care for the well-being of Native peoples whose way of life depends on the prey species. If people feel this way, they will experience a gain or loss of utility from a predator control program that must be accounted for in an economic analysis, just as gains or losses associated with impacts on recreation, food, or personal income are counted.

The value people place on an item for motives that are separate from their interest in using the item is known as existence value, nonuse value, or passive-use value. In many cases, people with no interest in using an item still believe it should be preserved and would be willing to pay something to ensure this. Their value will not be reflected in their use of the item, but it could be reflected in other forms of behavior, such as engaging in political or charitable activities, or engaging in a consumer boycott to express displeasure. Consequently, existence value cannot be discerned by analyzing people's demand to use the item by means of conventional market data. Existence value can, however, be estimated using a type of survey known as contingent valuation.

In the case of predator control in Alaska, evidence strongly suggests that existence values are important, at least for some segments of the resident and nonresident populations. The widespread public support for wolf protection and restoration in the lower 48 states and a general perception of wolves and brown bears as imperiled species, as well as the public controversy about wolf control within Alaska, point to this conclusion. However, the magnitude of the existence values, for both prey and predator species, which includes concern by non-Natives for the well-being of Native peoples in the absence of predator control, is unknown. No contingent valuation study on existence values associated with predator control has been conducted in Alaska.

Depending on the circumstances and the method of implementation, a boycott of tourism in Alaska is likely if a wolf control program were again authorized. Because the situation is so variable, it is impossible to predict with precision the extent of a future boycott or the magnitude of its effect on the state's economy. In addition to any loss of personal income from changes in tourism and any loss of utility for residents or nonresident associated with reduction in predator populations, there are costs to government agencies and others associated with planning and implementing a predator control program.

When calculating such costs, the relevant concept is the increment in cost due to the program, that is, its marginal cost. In this context opportunity costs may be more significant than direct costs. For an agency with a relatively fixed budget and staff, part of the incremental cost of providing a new activity may be the other services that have to be reduced or postponed when agency personnel are diverted to work on the new activity. To the extent that a program creates troublesome precedents for an agency or limits its future range of options, these too are opportunity costs associated with the program. Thus, the opportunity cost of the activity can be more significant than the direct costs.

The committee summarizes and synthesizes its findings in the following conclusions, which are presented in brief below and explained at greater length in chapter 8.

CONCLUSIONS AND RECOMMENDATIONS

Biological Issues

Conclusion 1: Wolves and bears in combination can limit prey populations.

There is clear evidence that wolves and bears can, under certain conditions, keep moose and caribou populations suppressed for many years, but evidence is insufficient to establish the existence of dual stable states, one of which has high densities of both predators and prey.

Conclusion 2: Wolf control has resulted in prey increases only when wolves were seriously reduced over a large area for at least four years.

Recommendation: Wolves and bears should be managed using an "adaptive management" approach in which management actions are planned so that it is possible to assess their outcome. That way managers can learn from the experience and avoid actions with uninterpretable outcomes or low probability of achieving their stated goals. Management agencies should be given the resources to conduct their management projects as basic research.

Conclusion 3: Expectations that managed populations in Alaska will remain stable are not justified.

Recommendation: Management objectives aimed at achieving stable populations of wolves, bears, and their prey should recognize that fluctuations in populations can be expected and provisions made for them in management plans. Before any predator management efforts are undertaken, the status of the predator and prey populations should be evaluated (including whether they are increasing or decreasing), and the carrying capacity of the prey's environment should be evaluated.

Conclusion 4: Data on habitat quality are inadequate.

Recommendation: ADFG should broaden the scope of their studies of both predator and prey species. They should collect better data on habitat quality and on bear ecology. They should continue to increase their development of long-term data sets. Three areas exist for which additional data are especially needed: bear foraging and population ecology, quantitative and qualitative changes in habitats, and the long-term consequences of predator control. Future research on these topics needs to be coordinated among the agencies that share jurisdictional authority over wildlife and wildlife habitats. The use of controlled fire should be further investigated as a tool for increasing the carrying capacity of moose habitat.

Conclusion 5: Modeling of population dynamics will enhance use of data already collected and enable more efficient use of limited resources.

Recommendation: Collaborative relationships among ADFG and the land management agencies and jurisdictions should be strengthened so that habitat studies and habitat management efforts are well-coordinated.

Conclusion 6: Wolves, bears, and their prey are vulnerable to human actions but in different ways.

Recommendation: Wildlife policy makers in Alaska should be more sensitive to signs of overharvest and more conservative in setting hunting regulations and designing control efforts, particularly with moose, caribou and bears.

Conclusion 7: The design of most past experiments and the data collected do not allow firm conclusions about whether wolf and bear reductions caused an increase in prey populations that lasted long after predator control ceased.

Recommendation: Future experiments should be based on more thorough assessment of baseline conditions and should be designed so the causes of subsequent population changes can be determined.

Conclusion 8: Perfect prediction is unattainable.

Conclusion 9: Many past predator control and management activities have been insufficiently monitored.

Recommendation: All control activities should be viewed as experiments with clear predictions. Control activities should be designed to include clearly specified monitoring protocols of sufficient duration to enable determination of whether the predictions are borne out and why.

Economic Aspects of Predator Management

Conclusion 10: Benefit-cost analyses of management changes require at least three categories of information: biological relationships among predators, prey, and their environment; human behavioral response to changes in perceived quality of the use in question (for example, hunting success), and frameworks for valuing the change in use (or availability) of the resource.

Conclusion 11: Evaluations of Alaska predator control programs have not gathered, analyzed, and assessed the full economic costs and benefits.

Recommendation: ADFG should increase its efforts to evaluate human responses to management actions at spatial and temporal scales sufficiently large to match the scale of the affected market. Travel cost models and contingent valuation should be applied to past and future management actions to improve assessments of value.

Conclusion 12: Social science research in Alaska is needed to support the design and evaluation of predator control experiments.

Recommendation: A formal procedure should be created, with adequate resources and trained personnel, to gather relevant economic, social, and cultural data, and incorporate this information into management and decision-making processes. The specific tools of benefit-cost analysis and applied anthropology should be used in the analyses performed on those data.

Conclusion 13: Wildlife is, by definition, a public resource.

Recommendation: Procedures should be developed to allow the public to be substantively involved at all stages of both the policy and regulatory process.

Conclusion 14: Greater potential for agreement may exist among Alaska's diverse constituency than is generally assumed.

Recommendation: ADFG and the Alaska public should engage in the development of a long-term strategic plan for the State's wildlife resources that is periodically revised, as necessary.

Conclusion 15: Conflicts over management and control of predators are likely to continue indefinitely.

Recommendation: A formal conflict resolution process should be developed and adopted to help avoid the kind of intractable and wasteful dispute that has characterized the recent history of wolf and bear management in Alaska.

Conclusion 16: Decentralization of decision-making is not a panacea for solving wildlife management problems, but is likely to be helpful in many circumstances, particularly in rural communities.

Recommendation: Decision-making should be partly decentralized through formal consultation procedures where the views of local groups are solicited before decisions are made. In management situations involving rural and indigenous groups, more refined co-management decision-making structures should be developed where appropriate.

Conclusion 17: Interagency cooperation could improve management, reduce public confusion, and eliminate unnecessary duplication.

Recommendation: ADFG should assume a leadership role in strengthening cooperative agreements between the various jurisdictions and agencies involved in wolf and bear management in Alaska.

1

Introduction

BACKGROUND

The governor of Alaska, Tony Knowles, suspended the state's wolf control program in late 1994 shortly after taking office because he judged it to be unacceptable in its treatment of both wolves and nontargeted species. He stated that he would not reinstate the predator control program unless it met 3 tests: it must be based on solid science, a full cost-benefit analysis must show that the program makes economic sense for Alaskans, and it must have broad public support. In July 1995, Governor Knowles asked the National Academy of Sciences to undertake a scientific and economic review of the management of wolves and bears in Alaska.

Management to enhance moose and caribou populations in Alaska typically includes management of wolves and bears. The essence of wolf control is that wolf populations are reduced, usually by killing wolves, so that more moose and caribou are available for human use. Bears also compete with people for moose and caribou, although they have received considerably less attention than wolves.

The committee's task was difficult because the scientific and economic issues are embedded in a complex political and ethical context. Lifestyles and cultures in Alaska range from those of people who depend almost completely on wildlife resources to those of people who buy all their food at supermarkets. Alaskans exhibit the full array of values about natural resources, but Alaskans as a group are more knowledgeable about wolves and more committed to coexisting with them than are people elsewhere (Kellert 1985). At the same time, Alaskans have more utilitarian and less preservationist attitudes toward wolves than do

14

people elsewhere in the United States. Owing to the large amount of federally managed land in Alaska and its great expanses of wilderness, people living outside Alaska have strong interests in resource management in the state, and they believe that they have a right to influence wildlife management policies in Alaska. Given that Alaska's tourism industry is second only to fishing in number of people employed, the presence of outsiders is an unavoidable fact of today's world; but many Alaskans resent it.

Scientific and economic approaches cannot resolve ethical and esthetic disagreements about predator control and management. However, management policies are more likely to be acceptable if the public has confidence in the scientific and socioeconomic analyses on which management is based than if the underlying science and economic analyses are in dispute. Once the distraction of disputed science is eliminated, conflicts in wildlife management that go beyond scientific issues can be better clarified, an essential step toward ensuring broad public support.

History of Wildlife Management in Alaska

Aboriginal Alaskans have hunted and trapped wolves for thousands of years. The extent of their past influence on wolf populations is unknown. During the early part of the 20th century, both the government and private parties conducted indiscriminate wolf control. Bounties were paid for wolves. During the 1950s, the federal government conducted systematic wolf control with poison and aerial shooting, which reduced wolf numbers in many parts of Alaska.

When Alaska became a state in 1959, the newly formed Alaska Department of Fish and Game (ADFG) suspended all wolf control programs and reclassified wolves as big game animals and furbearers. The state was divided into 26 game management units (GMUs), and unit-based management was initiated (figure 1.1). In the late 1960s and early 1970s, a series of severe winters coincided with high numbers of wolves and bears and, in some areas, very high harvests of moose and caribou by people. As a result, ungulate populations declined rapidly over much of Alaska. In response, ADFG reduced or eliminated hunting of moose and caribou, embarked on a program of fire management to improve habitats, and conducted limited wolf control programs. In some areas, these programs appeared to restore prey populations. The programs were not controversial at the time, but attitudes toward wildlife have changed since then.

Current Wildlife Management in Alaska

Wildlife management policies in Alaska are set by the Board of Game (BOG) whose members are appointed by the governor. Management of wildlife in Alaska has been carried out by ADFG's Division of Wildlife Conservation, although management of subsistence use of wildlife has undergone change in re-

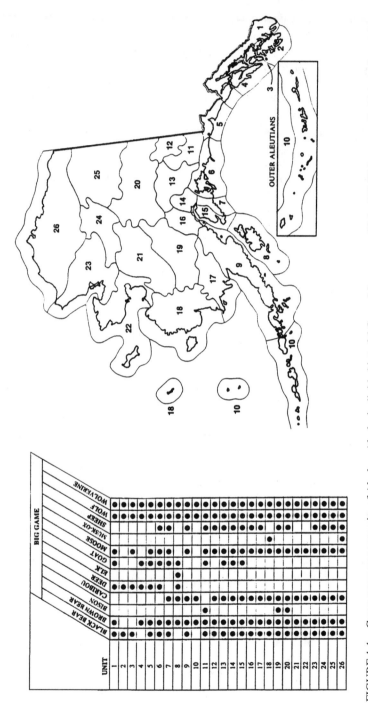

FIGURE 1.1 Game management units of Alaska. Alaska is divided into 26 Game Management Units (GMUs), many of which are larger than entire states. For example, GMU 20 is larger than the state of New York. ADFG has conducted wolf control programs in GMUs 13, 19, 20, and 21. Wolf control programs in GMUs 13 and 20 are discussed in Chapter 5. The dots in the table indicate where big game species can be hunted.

cent years. The commissioner of fish and game, who is appointed by the governor, heads ADFG and is designated an ex officio member of BOG. The commission also presents regulatory and policy proposals to BOG for consideration and approval. ADFG is responsible for implementing policies approved by BOG. BOG has constitutional and statutory authority to set game policy direction and to regulate game harvest and management on the basis of proposals by ADFG and the public, but it has no fiscal authority. ADFG prepares the budget for its management practices which is subject to the Governor's approval as well as that of the legislature. Nevertheless, ADFG has considerable latitude in adopting and implementing BOG regulations and policies. Not surprisingly, relationships between the 2 bodies under these conditions are sometimes contentious.

In 1989, state management of wildlife for subsistence use was determined to be out of compliance with federal subsistence requirements. Since then, a Federal Subsistence Board has been established to develop regulations for subsistence use of wildlife on federal public lands in Alaska. BOG retains authority for regulating subsistence use of wildlife on state and private lands.

Most management actions implemented by ADFG are not controversial, but wolf control and management in Alaska have become increasingly difficult because public perceptions of predators' roles and values and of the ethics of predator control are changing. Conflicts among people with different interests in wolves are intense. As a result, groups that could be united in ensuring the future well-being of wildlife populations regularly work at cross purposes.

The first concerted effort to resolve the conflict over wolf management began in 1988 when the Division of Wildlife Conservation decided to attempt to develop a broadly acceptable management plan with full and effective public participation. The division opened communication with a wide array of groups and individuals to discuss wildlife management. The Wolf Management Planning Team was formed in 1990; its 12 members were selected by the Division of Wildlife Conservation from among 70 nominees proposed by all groups interested in wildlife management. The team members represented rural and urban hunting, guiding, trapping, ecotourism and wildlife protection interests. A representative from a national conservation group and the deputy director of the Division of Wildlife Conservation also served on the team. All members shared the goal of maintaining viable wolf populations throughout the state, and they indicated a willingness to work with others to find acceptable solutions. A member of BOG monitored and advised the panel.

The team met monthly from November 1990 through April 1991 and submitted its final report to the Division of Wildlife Conservation in June 1991. The final report was based on a detailed review of information about wolf biology, predator-prey relationships, population dynamics, past control efforts, and hunting and trapping statistics. The team held 2 public forums and considered a wide range of management options. The report articulated goals, principles, and strategies for habitat conservation and set forth consumptive (hunting and trapping) and nonconsumptive (ecotourism and wildlife viewing) uses of wolves that its members believed should guide all management plans. The key management strategy was a system of 6 zones that varied from areas where no wolves would be killed to areas of intensive management where wolf control would be allowed. The report also outlined the conditions under which wolf control should be carried out.

The division then drafted a strategic wolf management plan that incorporated nearly all the team's recommendations. The draft plan was released to the public in September 1991, and it was discussed in numerous public meetings throughout Alaska. In late October 1991, the planning team and BOG held a 2-day workshop to review the plan for its consistency with the team's recommendations.

After a 2-day public hearing, BOG invited 2 members of the planning team and several members of the public to join in a detailed revision of the plan. The revised plan, which was unanimously adopted by BOG, required the Division of Wildlife Conservation to prepare area-specific management plans for all of Alaska. Priority was given to preparing plans for game management units 11, 12, 13, 14, and 20—areas where wolves were believed to be substantially suppressing prey populations. Area-specific plans were prepared, reviewed, revised, and finally adopted by BOG in November 1992. The board also adopted 3 implementation plans for areas where wolf control could occur in 1993.

At that time, many groups with different interests were willing to accept the compromise proposals which allowed for limited wolf control, and it appeared that consensus among the major interests in the state had been reached. But the apparent consensus was to be short-lived. Some environmentalists who had participated in the planning process and agreed to limited wolf control felt that BOG, in making its final determinations in a closed session, did not appreciate the magnitude of their concessions. Also, BOG underestimated the strength of concerns of people living outside Alaska.

Release of the board's plans triggered national and international attention. Some of the information that was publicized about the scope and intent of the program was inaccurate or misleading, and many people outside Alaska were not sufficiently informed to distinguish between accurate and misleading information. Whatever the accuracy of the portrayals of the wolf control program, the result was a call for a national boycott of tourism in Alaska, organized by a coalition of animal rights groups. The resulting social and economic pressures were so great that on December 4, 1992, Governor Walter Hickel suspended all wolf control pending a summit meeting in Fairbanks to which conservation, environmental, and animal rights groups were invited. On December 22, the commissioner of fish and game suspended all aerial wolf control in 1993—an action that stopped the tourism boycott.

About 1,500 people attended the summit, which was held January 16-18, 1993. Professional facilitators were hired to conduct the meeting. Speakers reviewed wolf natural history, predator-prey dynamics, the planning process, effects on tourism, and Alaskan, national, and international perspectives on wolf management. Participants divided into 9 groups facilitated by representatives of the Office of the Ombudsman to discuss the planning process and management decisions. From the lists of consensus points compiled by the groups, the facilitators identified the following issues of agreement:

- The planning process used, especially that involving the Alaska Wolf Management Planning Team, was good, and ADFG need not start over.
- ADFG should begin active education and information efforts about wolves.
- The Wolf Specialist Guidelines of the International Union for Conservation of Nature and Natural Resources (IUCN 1983) regarding wolf control and some elements of the Yukon Territory's wolf management plan (Yukon Renewable Resources 1992) should be incorporated into Alaska's wolf management plan.
- The state needed to take steps to make its BOG and advisory committee process more broadly representative of the public's diverse interests.
- More time was required for a fair and open public process to be successful.

All 9 groups concluded that wolf control might be appropriate under some conditions, but the short time available prevented them from reaching consensus on when and how.

On January 26, 1993, BOG repealed all area-specific plans and amended the strategic wolf plan to delete all references to zone management. The Division of Wildlife Conservation was instructed to prepare new regulation proposals that would give the board greater flexibility in addressing wolf control and management. The Division of Wildlife Conservation drafted 2 proposals: one to deal with the chronically low numbers of the Fortymile caribou herd; the other to deal with the continuing decline of the Delta caribou herd. At its meeting in June 1993, the board authorized a 3-year experimental ground-based control program for game management unit 20A. Its goals were to reverse the decline of the Delta caribou herd, to provide a harvest of 600-800 caribou by 1996, and to determine the effectiveness of ground-based wolf control.

ADFG initiated its wolf control program in GMU 20A during the autumn of 1993. In April 1994, it announced that 98 wolves had been trapped by the agency and another 52-54 by private trappers in GMU 20A. Wolf control was reinitiated during the autumn of 1994 but was abruptly terminated by Governor Knowles in response to an event on November 29, 1994.

On November 28, 1994, a biologist hired by animal rights groups to monitor the state's wolf trapping program flew over GMU 20A and observed what he believed to be a dead wolf in an ADFG snare. He returned by helicopter the next day, accompanied by 2 reporters on the *Anchorage Daily News* staff. They landed and located 4 wolves in snares; only one was dead. Several hours later, an ADFG agent arrived at the scene to dispatch the wolves, one of which had mutilated itself. Because the agent used improper ammunition, 5 shots were required to kill the first wolf. The incident was filmed and later was aired widely on television in the United States where it precipitated a powerful negative reaction across the nation.

THE COMMITTEE AND ITS MANDATE

In July of 1995, Governor Knowles wrote to the National Academy of Sciences requesting that it consider undertaking this study. He noted that the issues surrounding control and management of predators to increase the harvest of moose and caribou are of great interest not only to Alaskans, but to people across the country (see appendix A). Acknowledging that there will always be disagreement about predator control from ethical and other perspectives, he expressed hope that public confidence in the science and socioeconomics on which management is based would go a long way toward public acceptance of a management program.

The National Research Council (NRC) is the operating agency of the National Academy of Sciences. The NRC's Commission on Life Sciences' Board on Biology appointed the Committee on Management of Wolf and Bear Populations in Alaska. The committee members include people with expertise in ecology, population biology, mammalogy, wildlife management and policy, and economics (see appendix B). Several of the committee members are Alaskans; others have worked in Alaska on wildlife biology or economic issues. Members were appointed not only on the basis of their expertise, but also to provide a variety of perspectives. As with all NRC committees, committee members served as individuals and not as representatives of particular organizations, and the study was conducted independently of the sponsor, the Alaska Department of Fish and Game.

The committee was asked to assess Alaskan "wolf management and control and bear management." ADFG defines these terms in very specific ways and these definitions are used by the committee in this report (see box).

The mandate to the committee was to address wolf and bear management from both biological and economic perspectives, and to address each of the questions listed below:

Biological Aspects

• To what extent do existing research and management data provide a sound scientific basis for wolf control and brown bear management in Alaska?

• To what extent does current knowledge allow accurate prediction of the effect of a predator control program on predator and prey populations?

Definitions used by ADFG

Wolf control—a program defined in regulation (5AAC 92.110) by the Board of Game designed to reduce a wolf population to a specified level in a specific geographic area for a designated period of time to increase one or more prey species to meet population objectives for those species as adopted by the Board in regulation.

Wolf management—establishment on an area basis of population goals, objectives, activities, and regulations (seasons, bag limits, and methods and means of take) consistent with the principles of sustained yield to ensure the long-term viability of a wolf population.

Brown/grizzly bear management—establishment on an area basis of population goals, objectives, activities, and regulations (seasons, bag limits, and methods and means of take) consistent with the principles of sustained yield to ensure the long-term viability of a brown/grizzly bear population

Source: Wayne Regelin, Division of Wildlife Conservation, ADFG

• What critical data gaps exist in scientific understanding about predator and prey populations, and what would be needed to fill them?

Economic Aspects

• What existing economic studies and economic research methods can be used to evaluate the economic costs and benefits of wolf and bear management?
• What additional economic methods or data would be necessary for a comprehensive assessment of the economic costs and benefits of such management and control programs?
• What strengths and limitations should be considered in assessing these economic analyses?

The committee was also asked to synthesize current biological, ecological, and behavioral knowledge relevant to the assessment and implementation of wolf control and brown bear management in Alaska, including an analysis of how accurately one can predict the effect of wolf control and brown bear management policies on predator and prey populations; to assess how economic analysis could be used to predict the costs and benefits of wolf control in Alaska; and to address issues in decision-making and develop a decision-making framework that incorporates the results of the biological and economic review. This included an assessment of the factors that need to be taken into account when evaluating the utility of predator control and management programs from both scientific and economic perspectives.

HOW THE COMMITTEE CARRIED OUT ITS TASK

The overall goal of this report is to assist the state of Alaska in its efforts to use the best available scientific and economic data in making and implementing wildlife management decisions. The committee recognizes that wildlife biologists and managers in Alaska must base their decisions on information gathered from a vast land, much of which is inaccessible by road. Inevitably, data on populations of large mammals in many parts of the Alaskan landscape are sparse. This situation is likely to persist indefinitely because the financial and human resources needed to gather detailed population data on large mammals in the entire state are unobtainable.

Therefore, the committee's most important task was to determine the degree to which additional information would be likely to change a management decision, to improve the ability to predict the outcome of implementing a management program, and to increase public confidence that decision-making was using all available relevant information and that the information was analyzed and interpreted appropriately. In other words, the committee attempted to assess the marginal value of additional information from the perspective of the uses to which the information would be put.

Scientists and managers generally use the same information in different ways because they make different kinds of decisions. Typically, scientists demand a high degree of certainty before they accept the validity of a hypothesis or theory; that is, without convincing data they are reluctant simply to assume that a hypothesis is true. Such caution is appropriate because acceptance of false hypotheses can cause serious disruptions in the progress of scientific knowledge; moreover, there usually is no urgency in reaching conclusions. However, if management decisions are delayed while additional scientific data are gathered, a management policy has, de facto, been established.

For example, in deciding how much information is needed before a decision is made to reduce or not to reduce a predator population, managers face the following dilemma. If they adopt very high standards of data completeness before initiating any action, they allow existing conditions to continue. If population reduction would be effective under the circumstances, they risk allowing the current situation to deteriorate. But if they initiate a reduction program based on relatively sparse data, they risk wasting scarce financial and human resources on an ineffective management option and eroding public confidence in the management agency.

Therefore, in its assessment of the adequacy of scientific, sociological, and economic knowledge supporting control and management actions, the committee attempted to determine the likelihood that additional information would lead to different management decisions. In addition, we asked which information collected during a management program would be most valuable in removing uncertainties that have important policy implications. For example, if a control program is initiated when the available information is judged to be barely sufficient to justify the decision, what information should be gathered during the program to enable a manager to decide in a timely manner to terminate or continue the program? Should the same information be gathered if a decision to control was based on extensive data?

Because the control and management of wolves and bears in Alaska encompass a broad array of scientific and economic issues that vary geographically, the committee used a variety of methods to assemble the information that it needed to carry out its mandate. The committee held 4 meetings in the period from September 1996 to March 1997, 3 in Seattle and 1 extended meeting in Alaska that included forums in Anchorage, Fairbanks, and 3 villages. Before the first meeting of the committee, letters were sent to the full array of organizations concerned with wolf and bear management in Alaska. The committee solicited responses from organizations in and outside Alaska, from organizations opposed to wolf control, from proponents of wolf control, and from those with no formal position but whose members might be influenced by different management policies. The organizations included tourism associations and associations of sport hunters, Alaskan Natives, trappers, and animal welfare advocates. All recipients of letters were invited to submit written statements containing their views about the most appropriate and most feasible goals for predator control and management in Alaska. They were asked to identify and, if possible, provide information that the committee needed to analyze. All recipients were encouraged to pass the letters of invitation to other persons and groups that might have information useful to the committee.

To cast its information-gathering net even more broadly, the committee established a site on the World Wide Web where information about the study was posted and where interested parties were invited to submit data. Notices inviting members of the general public to present information to the committee were placed in the Anchorage and Fairbanks newspapers.

During its meetings in Alaska on October 23-26, 1996, the committee met with members of the Alaska Board of Game, and with wildlife scientists, managers, and social scientists employed by the Alaska Department of Fish and Game, the US Fish and Wildlife Service, the National Biological Service (now the Biological Resources Division of the US Geological Survey), and the University of Alaska. It also met with independent scientists. Committee members flew to and discussed predator management and control with Alaskan Native leaders and local residents in Aniak, Galena, and McGrath. During public sessions in Anchorage and Fairbanks, the committee received oral presentations and written statements from a wide variety of concerned people.

The committee also compiled and analyzed scientific publications on relationships between wolves, bears, and their prey, and control and management programs in Alaska and northwestern Canada. At its first meeting, the committee determined that its efforts should be focused mainly on 5 mammal species. It selected the 3 predators: wolf (*Canis lupus*), brown bear (*Ursus arctos*), and American black bear (*Ursus americanus*). Of the various prey, it selected 2 for primary attention: caribou (*Rangifer tarandus*) and moose (*Alces alces*). Dall's sheep (*Ovis dalli*) and black-tailed deer (*Odocoileus hemionus*), which are also preyed on by wolves and bears, are not discussed in detail in this report because the goal of wolf and bear reductions typically has been to increase moose and caribou populations. Collectively, large-hooved herbivorous mammals—such as moose, caribou, deer, sheep and goats—are called ungulates, and this term is used throughout the report.

The committee analyzed all available information on the economic values of consumptive and nonconsumptive use of the focal species in Alaska and data on public attitudes toward predators, their prey, and the goals and methods used in predator management and control.

The most important categories of information used by the committee in preparing its report were the following:

- **The goals and objectives of wolf and bear control and management in Alaska as perceived by varied segments of the Alaskan public.** Such information is necessary because evaluations of the adequacy of information that supports control efforts must be tied to the objectives of the control programs.

- **Alaskan Natives' and other Alaskan residents' knowledge of predators and prey in Alaska, the spatial and temporal scales of their knowledge, and how such knowledge is preserved and communicated.** Information can be gathered and interpreted in different ways, and local and traditional knowledge is an important source of information about wild populations. Different segments of the Alaskan population have different relationships with wildlife populations, and, consequently, they have different perceptions of the need for and goals of predator control programs.

- **Information on trends in populations of predators and prey in managed and unmanaged areas.** Relationships between predators and prey constitute the fundamental data used by managers to determine when and where particular management or control efforts might be expected to accomplish their intended goals. There are few areas in Alaska where no management of predators and prey takes place, but the type and intensity of management efforts differ widely from area to area. Control efforts are typically initiated in areas where biologists and managers believe that predators are preventing prey populations from increasing, because in such situations predator control can possibly result in substantial increases in prey population densities.

- **Information on past control programs.** The data that were used by managers to justify the need for a control program, the conditions when the control program was launched, the type of control that was carried out, and the responses of both predators and prey to the control efforts are the major empirical data that the committee used to assess the adequacy of the scientific underpinning of management and control decisions.

  • **Attitudes of all segments of the Alaskan population toward predator management and control and the methods used to accomplish them**. Although the mandate to the committee did not include an assessment of public attitudes and values concerning predator management and control, public responses to control programs have important economic repercussions. Therefore, the committee needed this information for its task of assessing the economic implications of management decisions.

  • **The economic value of consumptive and nonconsumptive use of the target species**. Alaska residents and visitors interact with and value predators and their prey in many ways, not all of which can readily be quantified in monetary terms. To accomplish its mandate, the committee gathered information on the current economic value of wolves, bears, and their prey to different user groups. It then discusses how economic value of wildlife should be considered.

ORGANIZATION OF THE REPORT

In chapter 2, the committee provides a concise history of predator management and control in Alaska. This information establishes the attitudinal and institutional contexts in which current predator and prey management is carried out.

Then, in chapter 3, the committee first provides an overview of Alaskan biomes and habitats and the relevant life-history traits of the species of concern. That information gives the *ecological context* in which predator and prey management is carried out. Then, in chapter 4, the committee provides a summary of current theories about predator-prey interactions and how they have guided the implementation of predator control programs in Alaska and northern Canada. This constitutes the *conceptual context* guiding management decisions.

In chapter 5, the committee analyzes past predator control efforts in Alaska and northwestern Canada to determine the success of those efforts and the degree to which they provide a basis for predicting success or failure of future control efforts.

Chapter 6 reviews Alaska's current economy and the contributions of consumptive and nonconsumptive use of wildlife to the economic welfare of citizens of the state and then evaluates the economic implications of predator control programs and how they are carried out. A review of how economic information is used by Alaskan agencies in making wildlife management decisions is also provided.

Chapter 7 takes a closer look at decision-making in wildlife management and analyzes in greater detail the implications of the fact that such decisions must always be made under conditions of substantial scientific and social uncertainty.

Chapter 8 presents the committee's conclusions and recommendations. The committee emphasizes that in its recommendations it is not attempting to tell the state of Alaska whether to implement predator control or when and where control might be carried out. Indeed, the committee was not mandated to make such recommendations. Instead, its recommendations embody the committee's suggestions of how scientific, sociologic, and economic information should best be used to make such decisions. The recommendations also include the committee's suggestions as to the most-important information gaps that need to be filled if predator management and control decisions and implementation are to receive broad public support.

REFERENCES

IUCN (International Union for Conservation of Nature). 1983. 15th session of the General Assembly of IUCN and 15th IUCN technical meeting. Gland, Switzerland.

Kellert SR. 1985. Public perceptions of predators, particularly the wolf and coyote. Biol Cons 31:167-189.

Yukon Renewable Resources. 1992. The Yukon wolf conservation and management plan. Whitehorse. 17 Pp.

2

Predator Control and Management: Past and Present

HISTORY OF PREDATOR CONTROL IN ALASKA

Predator control and management in Alaska probably began when humans first crossed over the Bering land bridge from Asia into North America. Early management activities are described in stories that have been transmitted orally through the generations. Life in the Arctic was harsh, and starvation was not uncommon. Survival depended on available prey and hunting success. Great importance was attached to the accuracy of the essential features of those stories because they encapsulated much of what had been learned about the arctic environment and how to survive and prosper in it.

Predator Management Before European Contact

The indigenous Athabascan Indians—the Koyukon, Ingalik, Tanana, Kutchin, Tanaina, Ahtena, and Han people—lived a seminomadic life throughout the Alaskan interior. They depended on the fish and wildlife of the region for food, clothing, and other materials. In the summer, they caught salmon with nets in the major rivers and streams; in the winter they hunted mainly moose and caribou (Clark 1974, Huntington and Rearden 1993, Osgood 1936, Sullivan 1942). As recorded in their stories, the overriding objective of wildlife management of those people was to reduce predator populations to allow for growth or maintenance of strong prey populations. The advice that was transmitted from generation to generation about wolves, bears, eagles, and sea otters in the stories of many tribal groups was that "we always need to keep them down," and that "it's important to stay ahead of them." Bears and eagles were taken at every opportunity; the level of harvest would be described today as "generous." Wolf populations were addressed in a different manner. The stories of the tribes, clans, and bands tell that they knew the location of almost all the wolf dens in their traditional hunting areas. People regularly culled wolf cubs at their dens to reduce wolf numbers; this was known as denning. The quantitative effects of those control efforts on predator populations are impossible to assess, but it is conventional to assert that reductions in average predator populations were substantial.

The Arrival of Europeans and the Early 20th Century

The long-established relationship of humans to fish and wildlife resources of the region changed markedly with the establishment of the fur trade, primarily during the early to mid-1800s, followed by the influx of gold-seekers around the turn of the century. Within a few years after the Klondike gold discoveries of 1896, prospectors had fanned out throughout interior Alaska, where they prospected in the summer and often trapped animals in the winter, living primarily on the wild food that they could obtain. The establishment of transportation and trading centers at Fairbanks, Eagle, Circle, Tanana, McGrath, Galena, and other locations far from sources of domestic meat in Washington and Oregon established a market for game meat. In the winter, many prospectors, miners, and others used dog teams to hunt extensively throughout interior Alaska for moose, caribou, and mountain sheep to supply their own needs and the growing markets. The pressure on large mammals for food for people and their dogs, which were then the primary mode of winter transportation, and the comparable pressure on wolves, bears, and other furbearers to supply the fur market were widespread and intense (Anderson 1913). Poison was widely used by trappers to take wolves and other furbearers (Peterson and others 1984).

The dispersal of mining activities throughout much of Alaska was followed by major changes in wildlife habitat. Forests were cut to provide wood for construction of buildings, for mine timbers, and for fuel for cabins, houses, and power generation. The extent of wildfire from accidental causes greatly increased, and fires were deliberately started to remove the forest or shrub cover to facilitate prospecting (Lutz 1950, Palmer 1942). As a consequence of the intense human pressure on wildlife and the alteration of habitats through clearing of forests and extensive forest fires during the first half of this century, populations of moose, caribou, mountain sheep, wolves, and bears were reduced to historically low levels.

Low levels of game populations and the increased human pressure on them in the early part of this century were factors in the establishment in 1925 of the Alaska Game Commission which imposed the first game-harvest regulations. Wolves were generally viewed as competitors for the moose, caribou, and mountain sheep sought by humans, as well as being responsible for the low abundance of other furbearers. In 1915, the newly created Territorial Legislature established a bounty on wolves that remained in effect until after statehood. The reported annual harvest of wolves in Alaska, based largely on bounty payments, was less than 100 around the turn of the century, between 100 and 300 before the 1920s, and from 350 to more than 1,000 during 1923-1959. In spite of growing opposition nationally, wolf control, ostensibly to protect ungulate populations from depletion was periodically officially sanctioned until 1951, even in Mt. McKinley National Park (later to become Denali National Park).

Predator control in Alaska paralleled that elsewhere in the United States. Especially in the West, predator control has long been a major public priority and official public policy, as evidenced by laws that provided for bounties on many predator species and programs that were funded by the public. Bounties have at times been offered for predators from hawks and owls to brown bears and wolves. In Alaska, bald eagles and Dolly Varden char were at one time the subject of bounties.

Early Wildlife Management

When Alaska was a territory, management of wildlife was under the jurisdiction of the federal government, first through the Biological Survey, which later became the Bureau of Sport Fisheries and Wildlife and then the Fish and Wildlife Service (FWS). However, the Alaska Game Commission, made up of citizens of the territory, established methods and means of harvesting wildlife and set seasons and bag limits. In 1948, the FWS Branch of Predator and Rodent Control began operating in Alaska with the primary assignment of killing wolves and coyotes to bring about an increase in moose, caribou, deer, and mountain sheep. Primary control methods included the use of poisons, although year-round trapping of wolves was permitted. Strychnine, encapsulated in chunks of seal blubber, was scattered around carcasses of prey; and "coyote getters," made from cyanide-loaded cartridges that fired into the mouths of wolves or other carnivores that pulled on the scented capsules, were widely deployed. The federal control agency emphasized maximizing the total number of wolves and coyotes killed. Little effort was devoted to focus control in areas of presumed need.

The potential use of aircraft for wolf control became apparent as soon as light aircraft came into general use in Alaska in the 1930s. H.R. Ferguson, of Selawik, where reindeer were herded, commented in a letter to the Alaska Game Commission in 1936: "We are using a plane here carrying mail and see the wolves from the air and know if the Department would buy a light pusher plane and use a gunner with buckshot, this country would soon be cleaned up. It would be an unsportsman-like way to kill them, but it is the only way to hunt them." By the late 1940s, aerial hunting of wolves in winter from small aircraft had become an effective technique for killing wolves in open terrain. In 1952, FWS agents mounted an intensive aerial control campaign on Alaska's North Slope, killing more than 250 wolves in a few weeks by a combination of aerial shooting and poison baits. That control effort, in an area with abundant caribou, was based on the untested hypothesis that if caribou in the northern herds were allowed to increase beyond the capacity of their range they would migrate south into new ranges where they would become available to increased numbers of hunters (USFWS 1952).

Intensive wolf control by the FWS in Game Management Unit 13 (GMU 13), initiated in 1948, had reduced the wolf population to an estimated 13 by 1953 (Burkholder 1959). An increase in the Nelchina caribou herd from 4,000 in 1948 to 40,000 in 1955 followed (Bos 1975). By 1957, as the Nelchina caribou herd continued to increase, liberal caribou-hunting seasons and bag limits were established. Wolves were then offered complete protection in GMU 13 until 1967. That action was part of an experiment to assess the role of predation in the population dynamics of the Nelchina herd, and it was the first time wolves had been protected in Alaska outside of national parks.

Statehood

The Alaska Statehood Act was passed by Congress in 1959, and administration of Alaskan fish and wildlife resources was transferred to the Alaska Department of Fish and Game (ADFG) in 1960. The department was staffed by young, university-trained biologists who were aware of the changing attitudes toward wildlife management in North America. They were outspoken in their opposition to the use of poison to control predators, and its use was discontinued in 1960. The wolf was reclassified as a big-game animal and furbearer by the Board of Fish and Game (later split into the Board of Fisheries and the Board of Game) in 1963, but it was not until 1968 that the Alaskan legislature gave authority to the Board of Fish and Game to abolish bounties within game management units. By 1975, bounties were no longer paid on wolves except in southeastern Alaska.

During recent decades predator control has been attempted in only a relatively small part of the state and has been targeted to areas where it was believed that wolves and bears were keeping moose or caribou populations at very low levels. Bears are mainly managed as game animals, but management efforts are also directed at reducing danger to humans and eliminating bear attractants in areas settled by people.

DECISION-MAKING BY THE ALASKA DEPARTMENT OF FISH AND GAME

The Alaska Department of Fish and Game has developed a set of procedures for collecting data on wildlife populations, analyzing and interpreting the data, presenting its conclusions in a standardized format, and obtaining public comment on its assessments and proposed management decisions. Attempts are made to resolve conflicts over proposed management activities, but no formal conflict-resolution process has been implemented.

Data Collection and Assessment

Types of data collected by the ADFG Division of Wildlife Conservation vary with wildlife species and management objectives. Most data are collected every 1 or 2 years, but for some species, particularly those on which data are difficult and expensive to obtain or those whose management consequences are minimal, data are collected less often. In areas where substantial controversy is expected, such as areas where predator control is being considered, data collection is intensified.

The data are put into a standard format by area biologists (of whom there are 23), and the report is sent to the management coordinator for the region in which the data were collected (there are 4 regions). The coordinator reviews, evaluates, and edits the report, which is then released to the public as a federal-aid document. In a controversial case, the report might also be reviewed by research biologists. If an area biologist determines that a regulation change is needed to meet management objectives, he or she consults with the management coordinator, and a regulation proposal is prepared. Such a proposal is reviewed by other staff in the region and then sent to the headquarters in Juneau for staff review. If the proposal is approved by the ADFG director, it is published as a formal regulation proposal and submitted to the Board of Game (BOG).

The Division of Wildlife Conservation collects harvest data from hunters' reports and also conducts censuses of wolves and prey populations in each management area. In selected areas, they add to the census data by collecting information on calving rates and survival rates or body condition of prey. The data are used to assess the desirability of different management options. For instance, if a caribou herd exhibits high survival during years when there are few wolves per caribou and low survival during years when there are substantially more wolves per caribou, wolf control might be considered to be a valuable tool for increasing caribou populations. The analyses conducted range from verbal descriptions of data, through statistical analyses of correlations between wolves and prey survival, to specific demographic projections of caribou or moose populations based on assumptions of different mortality due to predators (before and after control). Most of the analyses designed to predict the likelihood that a proposed predator control program will increase prey populations are done by the wildlife biologists primarily responsible for collecting the data on the pertinent populations.

The ADFG Division of Subsistence has done a number of surveys on subsistence use in specific communities. In contrast to the Division of Wildlife Conservation, where the focus is on the biology of game animals, the Division of Subsistence surveys cover *all* subsistence resources used by a given community and the focus is on determining if the subsistence needs of communities are being met. Data collected by the Division of Subsistence have led to reforms in wildlife management policies, and local input has been an important factor in assessing wolf reduction plans. For example, the Division of Subsistence conducted a survey in the McGrath area to determine if subsistence needs were being met and if local people supported or opposed wolf reductions.

Public Involvement

Public attitudes are assessed only in very limited circumstances. Some surveys are being conducted to determine hunter satisfaction, and a limited program gathers social-science information about management of wildlife in the urban area of Anchorage. ADFG's public-involvement process consists primarily of BOG meetings and the Big Game Advisory Committee (BGAC) meetings that are held around the state each year. BOG holds 2 formal meetings each year in each region, during which all the regulations for that region are considered. This approach is designed to allow BOG to consider the multispecies effects when setting seasons and bag limits. In 1994, a human-dimensions program was started, but all financial support for this effort was eliminated from the ADFG budget during the last 2 legislative sessions. Little support is now received from either the legislative or the executive branch of the Alaska state government for social-science research.

There are 87 local BGACs around Alaska, each with 7-11 members. Most committees meet 3 or 4 times each year, although some, such as the committee for the Anchorage area, meet more often. ADFG spends more than $500,000 each year to support local BGACs. All BOG and BGAC meetings are open to the public. The amount of public participation at these meetings depends on the issues. ADFG publishes a regulation-proposal book before each meeting of BOG. The book, which may contain as many as 400 proposals, is distributed to the public through an extensive mailing list. Proposals may come from several sources, including local advisory committees, the public, and the ADFG staff. The public is encouraged to comment on the proposals in writing, and the comments are made available to BOG before its meetings.

The public can address any proposal on the table at the BOG meetings, but individual testimony is limited to 5 minutes. In situations with considerable public interest, additional public meetings and open houses are held before the BOG meetings. For predator control programs, ADFG is required to hold at least one public meeting. In practice it holds at least 3 meetings. In addition, the ADFG director and staff members meet regularly with the large wildlife organizations.

Conflict Resolution

ADFG has no formal conflict resolution process. In controversial situations, such as development of a wolf control plan, the department has experimented with outside conflict resolution through citizen-participation teams. In most cases, however, the conflicts are presented to BOG for resolution and decision. On some occasions, the public has involved the governor's office to seek resolution of conflicts between interest groups.

Strategic Planning and Policy-Making

No statewide strategic plans exist, and strategic planning processes are poorly developed. General operational plans were written in the middle 1970s, but they are out of date and are rarely used by the agency. The plans provided broad guidelines for management of species by geographic area. The major goals of the plans were maximizing harvest and consumptive opportunities. The public had little input into the development of the plans, and very few people even knew of their existence.

In contrast, many area-specific plans have been developed over the last 5 years. The plans are goal-oriented, and the public has been involved in their development. Most of the area plans were developed around a proposed predator control effort. Today, no standard process guides the development of area plans. Developed area plans are presented to BOG for its endorsement and use in establishing area-based regulations. Most area plans are centered around consumptive wildlife uses; there is little consideration in the plans for nonconsumptive use of game and nongame species.

In addition to area plans, ADFG has individual species-management plans to guide management; an example is the Muskox management plan. Where area plans or species plans have not been prepared, goals and objectives are set by ADFG to guide internal wildlife management decisions. The public is not involved in this process. Other than BOG and the local BGACs, the public is generally not aware of the goals and objectives.

REFERENCES

Anderson RM. 1913. Arctic game notes. Distribution of large game animals in the far north—extinction of the musk-ox—the chances for survival of moose and caribou, mountain sheep, polar bear and grizzly. Amer Mus Jour 13:5-21.

Bos GN. 1975. A partial analysis of the current population status of the Nelchina Caribou Herd. Pp. 170-180 *in* JR Luick, PC Lent, DR Klein, and RG White, Eds. Proc 1st International Reindeer and Caribou Symp Biol Pap Univ of Alaska, Spec Rep No. 1.

Burkholder BL. 1959. Movements and behavior of a wolf pack in Alaska. J Wildl Manage 23:1-11.

Clark AM. 1974. Koyukuk River culture. Mercury Series, Ethnology Service Paper, No. 18. Ottawa: National Museum of Man.

Huntington S and J Rearden. 1993. Shadows on the Koyukuk. Alaska Northwest Books, Seattle, WA. 235 Pp.

Lutz HJ. 1950. Ecological effects of forest fires in the interior of Alaska. Paper delivered to First Alaska Sci Conf, Washington, DC. 9 Pp. Yale School of Forestry, New Haven, CT

Osgood C. 1936. Contributions to the ethnology of the Kutchin. Yale Univ Pub in Anthro No. 7. Yale Univ Press, New Haven, CT.

Palmer LJ. 1942. Caribou versus fire in interior Alaska: a study of burned-over lichen ranges. Unpub Progress Report. US Fish and Wildlife Service, Juneau, AK. 13 Pp.

Peterson RO, J Woolington, and TN Bailey. 1984. Wolves of the Kenai Peninsula, Alaska. Wildl Monogr No. 88. 52 Pp.

Sullivan RJ. 1942. The Ten'a food quest. Catholic Univ of America, Anthropological Series No. 11. Catholic Univ of America Press, Washington, DC.

USFWS (US Fish and Wildlife Service). 1952. Predator control—Annual report. Dept Interior, US Fish and Wildlife Service, Alaska District, Juneau, AK. 39 Pp.

3

Alaska's People, Biomes, and Wildlife Species of Concern

INTRODUCTION

To evaluate a science-based program, one must understand the contexts in which the program is carried out, including the ecological, economic, and political contexts. This chapter provides an overview of the physical and biological environments in which wolves, bears, and their prey interact. The two following chapters discuss what is known about predator-prey interactions in general (chapter 4) and what has been learned from different attempts to alter predator-prey interactions through wolf and/or bear reductions (chapter 5). The socioeconomic environment is covered in chapter 6.

An overview of the climate, vegetation, and soils of Alaska and the major biomes into which Alaska has been divided serves 2 purposes. First, by illustrating the great diversity of Alaskan environments, it demonstrates why management programs must be based on area-specific information. Second, it shows how the vastness of Alaska and the limitations of personnel and financial resources available for biological research make it inevitable that management decisions are based on less-complete information than is desirable and than would be possible if resources were less limited or the area smaller.

This survey is followed by a review of the ecology and natural history of wolves, bears, and their primary prey—moose and caribou. The review demonstrates the substantial differences among the species of concern and provides a basis for designing management programs that are tailored to the life-history traits of the managed species and how they respond to changes in their environment.

THE PEOPLE OF ALASKA

The first humans in the Western Hemisphere are believed to have come from Asia across the Beringian land bridge into Alaska 12,000-15,000 years ago. The first to arrive were the Paleoindians, who spread throughout North America and South America and from whom most native American cultures derived, including the Haida and Tlingit Indians of the southeastern coast of Alaska (Greenberg 1987). Later migrations of people are believed responsible for the Athabascan Indian cultures that are present throughout the interior and south-central regions of Alaska and in parts of northwestern Canada. The marine-oriented Eskimos of Arctic, western, and southwestern Alaska (represented today by the Inupiat, Yup'ik, and Koniak cultures) arrived much later, apparently by boat across Bering Strait. The Aleut culture of the Aleutian Islands and adjacent Alaska Peninsula has its closest affinity to early Eskimo cultures.

Today, the human population of Alaska is about 610,000, with the majority concentrated in and around Anchorage, Fairbanks, Juneau, and smaller south-coastal cities of a few thousand each. The Alaskan population is younger than the rest of the United States (median age 30 years versus 33.4 years for the whole United States), and its rate of population increase in recent years is second only to that of Nevada (Alaska Bureau of Vital Statistics 1995). The nonindigenous residents of Alaska (those who are not Alaska Natives of Eskimo, Indian, or Aleut descent as defined by the Alaska Native Claims Settlement Act of 1971) make up about 84% of the Alaskan population, and about 80% of them live in urban communities. The non-Native residents of Alaska are primarily first- or second-generation immigrants from the other states and reflect the racial and ethnic diversity that characterizes the United States. There are some differences in the racial make-up between Alaska and the United States as a whole. Alaska's population is 4.1% black (12.1% for the entire United States), 16% Native American (0.8%), and 3.2% Hispanic of any race (9%), according to the 1990 US census. Alaska Natives currently make up 16.5% of the state's population and most live in rural communities (Wolfe 1996). There are about 225 rural communities of fewer than 500 residents scattered throughout the state but concentrated in southeastern Alaska. The residents of all but a few of those communities are predominantly Alaska Natives.

Human activities have had less effect on the ecosystems of Alaska than elsewhere in the United States. Conversion of land to agricultural use has been minimal, as is the extent of land alteration through mining and petroleum development. The greatest alteration of ecosystems has been through extensive logging of forests in southeastern Alaska. More than 40% of Alaska is managed by federal agencies through the National Park Service, Fish and Wildlife Service, and Bureau of Land Management.

BIOMES: CLIMATE, VEGETATION, SOILS, PERMAFROST

Alaska is one-fifth the size of the lower 48 states and occupies 1,477,270 km². It extends more than 20° in latitude from Pt. Barrow to Amatiguak Island in the Aleutians; it spans 42° in longitude from Portland Canal in southeastern Alaska to Attu Island in the western Aleutians. The topography, climate, and ecosystems of Alaska are characterized by great diversity (Selkregg 1976; Klein and others 1997). Alaska's coastline extends more than 54,700 km, bordering on the North Pacific, Bering Sea, and Arctic Ocean. The average length of the frost-free period varies from 40 days in the Arctic to more than 200 days in parts of southeastern Alaska. The Alaskan interior, because of its relatively warm, very long summer days, has relatively high plant productivity, except where permafrost is present. Permafrost is ground that remains perennially frozen except for a shallow summer-active layer. Annual precipitation ranges from less than 25 cm in the Arctic to 500 cm in parts of the Alexander Archipelago of southeastern Alaska. The low precipitation of interior Alaska would result in desert conditions at lower latitudes, but most of the winter precipitation remains on the land as snow until spring, and low summer evaporation rates and drainage (impeded by permafrost) retain soil moisture throughout the summer in most areas. Permafrost is present throughout most of the Arctic and northwestern Alaska except beneath lakes, rivers, and adjacent riparian zones. South of the Brooks Range in the interior and in southwestern Alaska, permafrost is discontinuous and confined mostly to lowlands, north-facing slopes, and higher elevations.

The position of Alaska between the cold Arctic Ocean and the relatively warm North Pacific, its extensive coastline and southern islands, and its high mountain ranges and associated ice fields with intervening and extensive lowlands are responsible for the ecological diversity that characterizes the state. Alaska can be divided into 6 major biogeographic regions or biomes (figure 3.1).

Coastal Temperate Rain Forest and Coast Range Mountains

The coastal temperate rain forest grows under the moderating maritime influence of relatively warm ocean currents of the North Pacific and Gulf of Alaska. It is a continuation of the temperate coniferous rain forest that extends from northern California along the northwestern coast to northern Kodiak Island. The distribution of large mammals in this region, because of its many islands and the relatively short time since it was heavily glaciated, is more complex than that in the remainder of Alaska. Sitka black-tailed deer (*Odocoileus hemionus sitkensis*) are common on the islands and adjacent mainland of southeastern Alaska and have been successfully introduced to the islands of Prince William Sound and to Kodiak Island. Moose (*Alces alces*) occupy the major river valleys that penetrate the Coast Range Mountains—such as the Stikine, Taku, and Chilkat rivers—but do not normally occur in the coniferous forests of the islands. Mountain goats

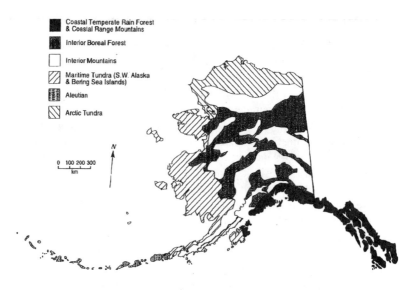

FIGURE 3.1   Major biogeographic regions of Alaska.

(*Oreamnos americanus*) occur in scattered localities in the mainland mountains and have been introduced to Baranof and Revilla islands. Brown, or grizzly, bears (*Ursus arctos*) are present on the mainland; Admiralty, Baranof, and Chichagof islands of southeastern Alaska; the islands of Prince William Sound; and Kodiak and Afognak islands. Black bears (*Ursus americanus*) and wolves (*Canis lupus*) are present throughout the mainland and on the islands of southeastern Alaska south of Frederick Sound. Densities of bears are high on the islands and mainland, especially in the vicinity of salmon spawning streams. Wolves vary in density in relation to the availability of their primary prey, blacktailed deer. Deer densities fluctuate with the frequency and severity of winters of deep snow. Deer habitat has been greatly modified by forest harvesting.

Interior Boreal Forest

Much of interior Alaska, which is sheltered by high mountains from the moist maritime air to the south and the cold Arctic air to the north, has a continental climate. Winters are cold and long; summers are warm and short. Seasonal changes are rapid. Altitude strongly influences plant growth, the presence and composition of forests, and the extent of permafrost. Fire, caused mostly by lightning, is a natural feature of the ecology of the interior boreal forest. The pattern of vegetation is a complex juxtaposition of plant communities that vary with fire history, soil temperatures, drainage, and exposure. The present boreal forest in interior Alaska is part of the northern boreal forest that extends from the

Atlantic coast of Canada across the northern portion of the continent into Alaska. Mammals characteristic of this major biome are similar throughout this vast area.

The extensive interior boreal forest region is broken by several mountain complexes that support typical mountain species, such as caribou (*Rangifer tarandus*) and brown bear. The lowland areas of mixed forest, shrub zones, and wetlands support moose and black bears. Moose also venture into the mountains, especially in summer. Wolves are present throughout the region, varying in density with the distribution and availability of prey, primarily moose and caribou. Mammal populations in this largely intact ecosystem undergo natural fluctuations, primarily in relation to variations in winter snow depth, plant succession after wildfire, and variation in rates of predation by wolves and bears.

Interior Mountains (Montane Habitats)

This biogeographic region is a complex of mountain ranges characterized by extreme physiographic variability. Elevation, slope steepness, and exposure vary widely locally and between major mountain masses. The distribution of vegetation, although dominated primarily by alpine forms, reflects the terrain variability, and vegetation at lower elevations includes elements from the boreal forest. Climatic variability is pronounced. Because oceanic air masses lose much of their moisture on the seaward sides of mountain ranges as rain or snow, interior sides are characterized by more arid conditions. The progression of summer plant growth is also highly variable, a condition that favors mammals—such as mountain sheep (*Ovis dalli*), moose, and caribou—that are able to move extensively over the variable terrain to forage. The depth of winter snows usually limits the availability of winter habitat for mountain sheep, moose, and caribou in these mountain areas. Brown bears and wolves are common, especially in areas of high prey densities.

Maritime Tundra (Southwestern Alaska and Bering Sea Islands)

The maritime tundra that dominates southwestern Alaska and the Bering Sea islands is the product of the cool climate generated by the cold Bering Sea waters. However, there is a gradation from the more-humid and milder conditions prevailing in Bristol Bay and the coastal Alaska peninsula bordering the Aleutian region to the Seward Peninsula, where the adjacent seas are ice-covered for 8 months of the year. Coastal wetlands are extensive throughout the region and dominate the broad expanse of the delta of the Yukon and Kuskokwim rivers (Selkregg 1976). In its climate and vegetation, the region is transitional between the Aleutian and Arctic biogeographic regions.

Moose, caribou, brown bears, and wolves have been absent or rare throughout much of this region in the past. In recent years, caribou re-entered the area from the expanding Mulchatna Herd, a re-established herd in the Kilbuck Mountains, and segments of the Western Arctic Herd, which winters on the western Seward Peninsula and in Kotzebue and Norton Sound drainages. About 30,000 reindeer (domesticated caribou introduced from Scandinavia) in several herds privately owned by Alaska Natives are grazed on the Seward Peninsula and have come into increasing conflict with wintering caribou from the Western Arctic Herd. Moose have become established in recent decades in riparian areas on the Seward Peninsula. Their populations appear to have peaked, with some possible declines, perhaps because the sparse winter habitats along the major rivers have become heavily browsed. Introduced muskoxen live on Nunivak Island, on Nelson Island on the Yukon Delta, and on the Seward Peninsula. Their expanding populations are hunted under permit systems by both subsistence and sport hunters. Brown bears are common in coastal areas of the Alaska Peninsula and present at lower densities on the Seward Peninsula.

Aleutian Region

The Aleutian Islands and adjacent Alaska Peninsula, an interface between the North Pacific and the Bering Sea, includes the southernmost land area in Alaska. The Aleutians, which extend nearly 1,900 km from the Alaska Peninsula to Attu Island, are renowned for their cool, foggy, and windy weather and temperature consistency. The mean daily temperature of 3.9°C has an annual range of only 9.4°C.

Native large terrestrial mammals are absent from virtually all the Aleutian Islands. In the Aleutian biogeographic region, only the southern portion of the Alaska Peninsula and closely adjacent Unimak Island support caribou (currently at a low density), brown bears, and wolves. The bears are moderately abundant and depend heavily on the productive salmon streams in the area. Wolf numbers are low, presumably because of the low prey density.

Arctic Tundra

The Brooks Range mountains separate the boreal forest and the Arctic tundra biogeographic regions. The Arctic tundra of Alaska, on the higher and drier ground at its southern limit, descends to the broad and wet coastal plain that continues to the Arctic Ocean. Coastal plain tundra is interspersed with thousands of shallow lakes. The Arctic tundra experiences strong northeasterly winds in winter that are generated by the Arctic high-pressure system over the frozen Arctic Ocean. The little snow that falls throughout the long winter is redistributed by winds into drifts wherever there is variation in the terrain.

In spite of climatic extremes of the Arctic tundra—which results in a short plant growth season, low mean annual temperature, and cold soils underlain by permafrost—this region is extremely productive of life, much of which (such as nesting birds and caribou) migrates out of the region during winter. The Arctic tundra supports Alaska's largest caribou herds, the Western Arctic Herd (about 500,000) and Porcupine Herd (about 170,000), as well as the smaller Central Arctic and Teshekpuk herds, each numbering around 20,000. In the latter half of the 20th century, moose moved into riparian habitats along the Collville, Sagavanirktok, Canning, and other larger rivers of the Arctic. Barren-ground brown bears occur at low densities, being more numerous in the southern foothills and western Arctic where they are important predators on young caribou during calving. Wolf densities are relatively low in the Arctic, largely because of the low density of ungulate prey in winter. Their numbers are highest in the northern foothills and adjacent Brooks Range mountains, where mountain sheep, moose, and some wintering caribou can be present. Muskoxen have been re-established in the Arctic through introductions in the 1960s and 1970s in the eastern and western areas. They are increasing and dispersing into unoccupied habitats. The total numbers in the Arctic tundra appear to be approaching 1,000.

ECOLOGY OF LARGE MAMMALS IN NORTHERN ECOSYSTEMS

The population dynamics of large mammals in southern Canada and the lower 48 states have been dramatically altered over the past two or three centuries. Not only have most of the natural habitats been converted to agricultural lands, managed forests, or high density human use areas, but dominant species such as wolves, grizzly bears, and millions of migratory bison have largely been eliminated. The population dynamics of large mammals in the pre-Columbian system are unknown. What remains of the system can be managed carefully to provide predictable harvests for hunters and trappers.

In northern Canada and Alaska, however, natural habitats have not been substantially altered. Alaskan ecosystems are still much the same as they were when Europeans first arrived in North America and caribou, moose, wolves, and bears are not threatened with extirpation. The large mammal system remains largely intact and is highly volatile. Most caribou migrate over large, and often unpredictable areas, and their numbers fluctuate enormously. For example, in Alaska, the Mulchatna herd increased from about 30,000 to 110,000 individuals between 1984 and 1993. In Quebec and Labrador, the George River herd increased from about 10,000 in the 1950s to approximately 800,000 in the 1990s. Such enormous shifts in one species influence the dynamics of the entire system, and large changes in moose abundance and distribution have also been evident. Biologists have debated how and when factors such as the very slow growth of terrestrial lichens, the highly variable Arctic weather, and predation at different caribou densities, influence caribou and moose populations. What isn't debated is that the northern systems are much more dynamic than what is left of the southern systems. Consequently, we cannot expect to manage northern systems to provide predictable numbers of ungulates and predictable harvests; Alaska is not Colorado, and caribou are not elk. Alaskans should not expect as constant a supply of game as people expect in the southern portion of the continent.

Many factors influence the population dynamics of moose, caribou, wolves, and bears. Among the most important are quantity and quality of food, weather, diseases, parasites, predation, intraspecific strife, and human harvest. These factors can act individually, but often they act in concert. For example, animals weakened by poor nutrition are more likely to succumb to disease or predators than well-fed animals are. Together, these factors determine the ecological carrying capacity of the environment—that is, the number of animals that can be supported on a long term basis by the resources of the environment (see box). Another definition of carrying capacity is the number that will persist if disease organisms and predators are present. Because the goal of predator control and management in Alaska is to increase prey numbers, we use the former definition of carrying capacity in the report. Economic carrying capacity, the density of animals that will allow maximum sustained harvest, is always lower than ecological carrying capacity (Caughley 1976 in Krebs 1994). Carrying capacities fluctuate because abiotic and biotic environmental factors change. Assessing environmental carrying capacity is important because success or failure of management programs often depends on the environmental conditions under which they are carried out.

Ecological carrying capacities of Alaskan environments for ungulates are low because arctic, alpine, and subalpine soils are typically poor in nutrients. In combination with short growing seasons, this limits the potential quality and production of forage. Shallow, high-latitude soils underlain by permafrost are fragile and easily damaged and vegetation may recover slowly from overgrazing.

Arctic and sub-Arctic plants grow rapidly during the short growing season, and growth rates are higher during warm, sunny summers than during cloudy, cold summers (Klein 1970). Plant species are adapted to different soils, terrain, exposure, and water availability (Maessen and others 1983) and are distributed in a mosaic throughout Alaska (Maessen and others 1983). This mosaic distribution influences the foraging patterns of large herbivores and omnivorous bears. Large herbivores, in turn, can strongly affect plant succession, species composition, and productivity. High densities of large herbivores, such as moose and caribou, can cause both short-term declines in aboveground plant biomass and long-term declines in the quality of plants (Leader-Williams and others 1981). Lower biomass or poorer-quality plants (for example, plants that have been severely browsed) reduce the carrying capacity of the habitat for ungulates (Klein 1968).

Plants respond to being fed on by moose or caribou in a variety of ways. Light grazing can stimulate growth, and plants that grow beyond the reach of browsers can complete their annual growth more rapidly (Edenius 1993; Edenius and others 1993; Molvar and others 1993; Robbins and others 1987). Browsing and grazing can also stimulate chemical changes that alter plant palatability (Edenius 1993; Bryant and Kuropat 1980). Thus, plants and herbivores are

Carrying Capacity

Ecological carrying capacity is the number of organisms that the resources of the environment can sustain in a particular region (Pianka 1978, Sharkey 1970). Populations that exceed their carrying capacity must ultimately decline. Populations below their carrying capacity will tend to increase toward it. Carrying capacity itself can vary from year to year because the availability of resources for a population varies from year to year. Because carrying capacities are themselves dynamic, they are difficult to estimate (Dhondt 1988, Pulliam and Haddad 1994). The notion of carrying capacity is crucial in studying predator-prey systems because the effect of predator removal depends largely on how much below its ecological carrying capacity a prey population is. If prey are far below their carrying capacity and if predators are the cause, predator removal can be effective in enhancing the abundance of prey. In contrast, regardless of how many prey are killed by predators, if prey populations are close to their ecological carrying capacity, removal of predators will not produce a marked increase in prey abundance.

involved in a coevolutionary process of defense against and response to predation (for example, Coley and others 1985; Edenius 1993; Rosenthal and Janzen 1979).

Fire is a natural part of northern ecosystems and has a major impact on plant community organization and structure, and, hence, carrying capacity for ungulates (Viereck 1973; Zackrisson 1977). After a fire, short-lived and fast-growing deciduous shrubs and trees colonize the area, and long-lived and slow-growing species decline (Scotter 1967). Lichens recover from fire much more slowly than shrubs, but, over a long time period, fire rejuvenates lichen communities and helps to create the mosaic of habitats that is characteristic of Arctic and sub-Arctic environments (Miller 1980; Zackrisson 1977). Fires might have increased with settlement and have been blamed for the decline of some caribou herds (Bergerud 1974; Miller 1980). However, after about 200-300 years without a fire, lichens become senescent and grow more slowly (Klein 1982).

Because abundance and distribution of animal populations depends on the availability of suitable habitat, habitat management should be explored as a management tool, especially where studies of habitat and surrogate measures of habitat quality (responses of ungulates such as age at first reproduction, reproductive rates, and growth rates) indicate that habitat quality is depressing the growth of ungulate populations. If the data indicate that a population is limited by habitat (especially food), habitat management programs (such as burning or crushing of vegetation) might be effective. Habitat management is a long-term process and requires long-term commitment. It will not solve short-term "emergency" situations, but it can improve food availability and it might buffer populations during catastrophic events, such as deep persistent snow. Finally, because habitat management is generally socially acceptable, it offers the potential to decrease or avoid the use of controversial methods (such as killing predators) to increase ungulate populations.

Wolf Ecology

Distribution and Density

Wolves occur throughout the Northern Hemisphere wherever large ungulates occur, from about 20° N latitude to the polar ice pack. They are present in habitats ranging from deserts to tundra, and although they still occur as far south as Saudi Arabia and India, they are most common in more-northern areas of Alaska, Canada, and Russia. In Alaska, wolves are found over most of their historical range, occupying about 85% of the state's 1.5×10^6 km^2 (figure 3.2; Stephenson and others 1995). Wolves are absent from areas that they did not colonize after the last glacial recession, including the Aleutian, Kodiak, Admiralty, Baranof, and Chichigof islands.

Wolf densities vary geographically. In Alaska, wolf densities range from about 2 to 20/1,000 km^2; total numbers were estimated at 5,900-7,200 during the winter of 1989-90 (Stephenson and others 1995). In the Arctic, densities of wolf populations are often less than 5/1,000 km^2, but maximal midwinter wolf densities in southern populations often exceed 40/1,000 km^2 (Fuller 1989a).

On the basis of data from more than 20 intensive studies that measured total average ungulate biomass (often more than 1 ungulate species) and average wolf populations for a period of several years, variations in wolf density in all of North America seem to be strongly correlated with variations in ungulate biomass (Fuller 1989a, 1995; Messier 1985). The relationship between prey abundance and wolf numbers can vary in areas with migratory versus nonmigratory prey or where prey concentrate seasonally. However, all available data suggest that, unless artificially depressed by humans, wolf numbers are typically limited by ungulate numbers and availability. Food availability is the dominant natural factor that limits wolf abundance.

Although the correlation between wolf density and prey abundance is high, the ratio of ungulate biomass to number of wolves is highest for heavily exploited wolf populations (Ballard and others 1987; Peterson and others 1984) or newly protected populations (for example, Fritts and Mech 1981; Wydeven and others 1995) and lowest for unexploited populations (Mech 1986; Oosenbrug and Carbyn 1982) or those where ungulates are heavily harvested (Kolenosky 1972). In Alaska, lightly harvested wolf populations occurred at much higher densities per unit of ungulate availability than heavily harvested populations (Gasaway and others 1992).

Changes in wolf density in response to varying prey density have been documented by long-term studies in northeastern Minnesota (Mech 1977, 1986), Isle Royale (Peterson and Page 1988), and southwestern Quebec (Messier and

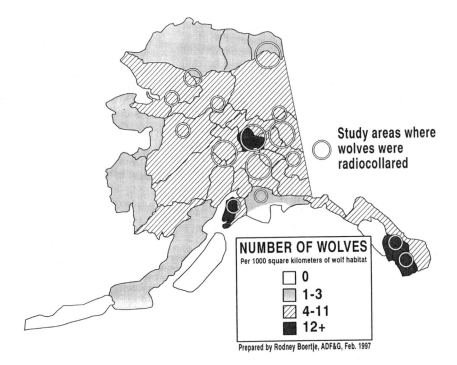

Study areas where
wyolves were
radiocollared

NUMBER OF WOLVES
Per 1000 square kilometers of wolf habitat

☐ **0**
▨ **1-3**
▨ **4-11**
■ **12+**

Prepared by Rodney Boertje, ADF&G, Feb. 1997

FIGURE 3.2   Alaska wolf population densities as estimated by Alaska Department of Fish and Game. Management biologists with aerial winter surveys and by contacting area trappers and pilots. Research biologists have radio-collared wolves from several packs in 11 Alaska study areas to verify estimates of wolf densities.

Crête 1985). Because the numerical response of an individual wolf population lags behind a change in prey density by up to several years, the ratio of ungulate biomass to wolves often differs among years and areas (Peterson and Page 1983).

Wolf densities can also vary with prey type. Population densities per unit of prey biomass are lower in areas where wolves prey mainly on moose than where they prey mainly on deer. The reason appears to be that moose are, on the average, less vulnerable to wolf predation (harder to catch) than are deer. Vulnerability of individual prey species may depend on which other prey species also are found in the area. For example, caribou are more vulnerable to wolf predation when they co-exist with moose, but moose and sheep are less vulnerable when caribou are present (Bergurud 1974, Dale and others 1995, Seip 1992).

In Alaska, wolf populations are estimated at different levels of precision, depending on management needs. The least-precise estimates are derived from a combination of information resulting from aerial surveys of wolf tracks, incidental observations, reports from the public, and sealing (mandatory registration) records. Such assessments often are made for a given game-management unit or subunit (for example, Hicks 1994) in the fall (pre-trapping season) and spring (post-trapping season) to make use of harvest data and to identify the annual population low that occurs in the early spring before pups are born. Aerial track surveys are conducted 1-3 days after a fresh snowfall (Stephenson 1975). Flight patterns concentrate on terrain on which tracks are visible and over which wolves are likely to travel. Wolf tracks are followed until wolves are sighted or until wolf numbers can be estimated from tracks. Such surveys provide an estimate of the minimal number of wolves in an area, and their accuracy depends on ease of observation and the intensity of search effort. Observations of wolf packs, by whatever method, that note location, pack size, and color composition, and harvest information obtained from sealing of wolf pelts can augment aerial-survey data and indicate the location of packs that were not detected by aerial surveys. These low-precision estimates are used throughout Alaska (except southeastern Alaska), where populations of neither wolves nor their prey are thought to be changing dramatically and where no management is contemplated in the near future.

The most-precise estimates come from censuses of radio-collared packs (as in Fuller and Snow 1988). During the last 25 years, nearly every well-studied wolf population has been studied by radio-marking one or more wolves in most of or all the packs in a study area of 1,000-5,000 km^2. During the winter, all packs are counted by radio-tracking and searching for tracks or wolves in areas where no packs are radio-collared but that are large enough to have a resident pack. Numbers of lone wolves are estimated by using the rate at which local pack members are known to disperse and settle, or by using rates determined for other populations (usually 10-15%). In Alaska, such intensive censuses have been conducted over a number of years in at least 9 areas where detailed predator-prey interaction studies were being conducted (figure 3.3).

Rigorous, statistically designed aerial track-sampling surveys have recently been conducted in several areas of Alaska (Ballard and others 1995; Becker and others in press). The method requires snow in which wolves leave continuous tracks, and a recent snow or windstorm is required to identify fresh tracks. Tracks are followed, and the size of the group making the tracks is identified. Comparisons with telemetry-derived census data have shown the technique to be accurate. It holds promise for making statistically valid estimates of wolf abundance over large areas that cannot be studied with radiotelemetry or do not merit the expense of more-intensive censuses.

Habitat and Diet

Four factors dominate wolf population dynamics: wolf density, ungulate density, human exploitation, and ungulate vulnerability (Fuller 1989a, 1995; Keith 1983). Rabies and other diseases or parasites might infrequently limit local wolf numbers, but no studies have suggested that they are a common limiting factor. Food habits of wolves have traditionally been assessed by analyzing scats (feces) during snow-free seasons and by finding kills in the winter. Although beavers, hares, and other small mammals can be seasonally important in some areas, they have not been shown to control either the distribution or the abundance of wolves. Vegetation type makes little difference to wolves as long as populations of hoofed prey are available. Prey availability is determined by both abundance and vulnerability. Deep snow or disease can make some prey more vulnerable and thus more "available" (Mech and others 1995). Availability of food ultimately affects nutritional levels and thus wolf reproduction, survival, and behavior.

Wolf populations are also influenced by hunting, trapping, and other control activities, but they seem little affected by snowmobiles, cars, trucks, logging, mining, and other human activities, except for accidental killing by humans or intentional killing where hunting or trapping for wolves or other species occurs.

Because wolves are great dispersers and can move to new areas fairly easily, wolf populations in marginal habitats (that is, where food resources are poor because of competition with humans or where there is high human-caused mortality) can be augmented by individuals from adjacent source or reservoir populations. Thus, the distance between populations plays an important role in wolf population ecology.

In sum, good wolf habitat is where ungulates are abundant and available, and where adverse human effects on wolves are low.

Social Structure and Movements

Packs originate when male and female wolves pair and produce pups. These wolves typically are dispersers from different packs that meet, travel together, find an "unoccupied" territory, and settle down. One of the wolves might have already settled in an area or might be the sole survivor of a pack with which a dispersing wolf of the opposite sex pairs. Packs vary in size from 2 to 25 or more. Average pack sizes, which usually range from 3 to 10, do not differ substantially among wolf populations whose major prey are different; that is, packs feeding mainly on black-tail deer are not usually larger than those feeding on caribou. Pack sizes also are similar at both high and low prey densities (but see Messier 1985).

Reproduction

Female wolves older than 22 months are capable of producing pups every year, but in a given year many potentially reproductive females in a pack do not produce pups. In particular, most wolf packs produce only 1 litter of pups per year (Packard and Mech 1980; Packard and others 1983), although 2 litters per pack have been reported (Ballard and others 1987; Harrington and others 1983;

Mech and Nelson 1989; Murie 1944; Peterson and others 1984; Van Ballenberghe 1983). If there are more than 2 female wolves of reproductive age in a pack, some do not breed. Consequently, populations with larger packs have a lower proportion of breeders (Ballard and others 1987; Peterson and others 1984). Peterson and others (1984) found that on the Kenai Peninsula in Alaska, increased harvest of wolves resulted in smaller packs and territories and establishment of new packs in vacated areas; as a result, breeders made up a higher proportion of the population, and the rate of pup production increased. It is not known why 2 females per pack produce pups in some areas more frequently than others (for example, Denali Park, Alaska). Pack sizes are not characteristically higher in these areas than in other geographically similar areas with similar prey bases.

In the 1960s, litter size was small in an unexploited population in Ontario (average, 4.9; Pimlott and others 1969) but large in exploited populations in Alaska (average, 6.5; Rausch 1969) and northeastern Minnesota (average, 6.4; Stenlund 1955). These reports led Van Ballenberghe and others (1975) and Keith (1983) to suggest that litter size might increase with greater ungulate biomass per wolf. More recent data from Alaska strongly confirm the assertion: litter sizes increased by 50% when available ungulate biomass per wolf increased by a factor of 6 (Boertje and Stephenson 1992).

Measurements of reproductive rates are based on examination of carcasses of adult females and, more rarely, observation of wild wolves (usually radio-collared) at den sites. The proportion of pups in early-winter populations is sometimes assessed by aerial observations (Peterson 1977).

Survival Rates and Causes of Mortality

Wolf survival is usually assessed in 2 ways. Indirectly, differences in consecutive population counts and known deaths due to "harvest" are used to estimate the proportion of wolves that have died. More directly, radio-collared wolves are monitored, and the rates at which they die are extrapolated to obtain an annual estimate of survival.

Survival of wolf pups in summer is difficult to measure, but data strongly suggest that survival is positively correlated with prey availability. Pup survival over the summer was almost double (89% compared with 48%) where per capita ungulate biomass was 4 times greater (Fuller 1989b). In northeastern Minnesota, pup survival decreased in years when the ungulate food base declined (Mech 1977; Seal and others 1975; Van Ballenberghe and Mech 1975). The percentage of pups in the population or in packs is highest in newly protected populations (Fritts and Mech 1981) and heavily exploited populations (Ballard and others 1987)—situations in which litters are large, pup survival is good, and ungulates are abundant (Harrington and others 1983; Keith 1974, 1983; Pimlott and others 1969).

Wolves die of a variety of natural factors, such as starvation, accidents, disease, and intraspecific strife. In populations where some human-caused deaths occur (and thus compensate for natural mortality), about 8% of individuals greater than 6 months of age can be lost each year (Ballard and others 1987; Fuller 1989a). On Isle Royale, where there are no human-caused deaths, annual mortality due to starvation and intraspecific strife (both related to relatively low food availability) ranged from 18% to 57% during a 20-year period (Peterson and Page 1988). Rabies, canine distemper, parvovirus, and such parasites as heartworm and sarcoptic mange, might be important causes of death, but documentation is scarce (Brand and others 1995).

Legally regulated killing of wolves can reduce or eliminate a population. Where focused wolf reduction programs have occurred, populations have been reduced by more than 60% in some years. In a few cases, site-specific control programs have eliminated entire packs (Fritts and others 1992). Even in some legally protected populations in Minnesota and Wisconsin, human-caused wolf deaths reached 20-30% per year (Berg and Kuehn 1982; Fritts and Mech 1981; Fuller 1989a; Wydeven 1993).

Rates of Population Change

Under favorable conditions, wolf populations can increase as rapidly as 50% per year (Hayes 1995). Potential rates of increase are higher after exploitation because increased per capita food availability results in increased pup production and survival (Keith 1983). Immigrations from surrounding areas could push the rate of increase even higher.

Given normal levels of natural mortality, sustainable harvest of wolves is probably less than 30% of the early winter population (Keith 1983). Wolf density on the Kenai Peninsula declined after 2 annual kills of greater than 40% but increased after harvests of less than 35% (Peterson and others 1984). Wolf populations remained stable after harvests of 16-24% of the early-winter population but declined by 20-52% after harvests of 42-61% of the population (Gasaway and others 1983). In the unexploited wolf population on Isle Royale in Michigan, annual mortality of adult-sized wolves was 13-16% when numbers were increasing or stable, 51% during a population crash, and 33-34% when the population later stabilized (Peterson and Page 1988). Wolf populations might be able to withstand harvests as high as 40% before the harvest causes a population decline. Determining the exact magnitude of harvest that will cause a wolf decline is difficult and depends on the methods of analysis used (Ballard and others 1987; Fuller 1989a).

Dispersal is a major means by which wolves expand their geographic range or immigrate to populations that could not otherwise sustain themselves. Radio-telemetry studies show that a wolf usually leaves home by itself, finds a mate or another pack that accepts it, settles down, and either begins its own pack or lives in the new pack (Fritts and Mech 1981; Rothman and Mech 1979). The proportion of dispersing lone wolves in a population varies seasonally, but a variety of studies have documented or estimated that lone wolves typically make up about 10-15% of a wolf population (Fuller 1989a).

Dispersing wolves typically establish territories or join packs within 50-100 km of the pack in which they were born (Fritts and Mech 1981; Fuller 1989a; Gese and Mech 1991; Wydeven and others 1995). However, dispersing wolves sometimes move longer distances. Fritts (1983) observed a wolf that traveled at least 550 km. Radio-collared wolves have traveled long distances between Wisconsin and Minnesota or Ontario (Mech and others 1994; Wydeven and others 1995), and a number of wolves have dispersed from western Minnesota far out onto the Dakota Plains (Licht and Fritts 1994). In Alaska, it is fairly common for wolves to disperse hundreds of kilometers from their natal range (figure 3.3; Stephenson and others 1995). Thus "connectivity" of populations is relatively high (Fritts and Carbyn 1995).

Consequences of Control on Wolf Populations

When wolves have been intentionally harvested heavily in 1 or more years during a control effort, their numbers have declined precipitously; but when control efforts have ceased, wolf numbers have rebounded to 88-112% of pre-control densities within 3-5 years (table 3.1). In south-central Alaska (GMU 13) wolf numbers increased from minimum post-control density by only 85%, but legal harvest was continued after control ended to keep wolf numbers from reaching pre-control density. Annual rates of increase during such recoveries, and others rates for which no corresponding pre-control densities were calculated, averaged about 26% and were the result of dispersal of young wolves into new territories, rapid pair bonding, territorial establishment and breeding, and shifts of pack ranges. During population recovery, survival rates of wolves are typically very high, dispersal rates are low, reproduction is normal, and packs are likely to split (for example, Hayes 1995). It has been suggested that harvest of wolves can actually result in an increase in the number of packs in an area, but there are few data to support this (but see Peterson and others 1984). Given relatively abundant prey and an annual increase of 26% per year, wolf populations potentially can recover to pre-control numbers in as little as 3 years. When wolves have recolonized areas from which they were extirpated, their populations have also increased rapidly (for example, Fritts and Mech 1981; Wydeven and others 1995).

Although wolf populations can recover numerically from major control efforts within a few years, it has been suggested that managers need to consider the special features of wolf social organization and related behavior (Haber 1996). Some critics of wolf control have argued that human harvest of ungulates is biologically justifiable because ungulates have evolved as prey but that it is not justifiable for humans to kill wolves because wolves are highly evolved social

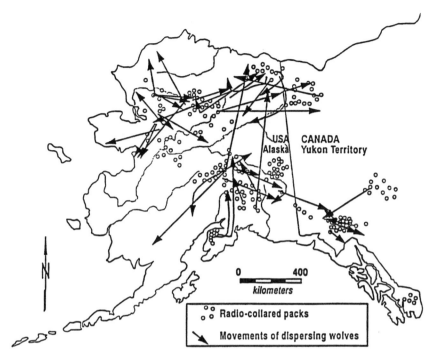

FIGURE 3.3   General location of radio-marked wolf packs. Wolf packs were studied in Alaska and the adjacent Yukon from 1975 to 1991, and known long-range movements of dispersing wolves. Packs were monitored for periods of 2 to 8 years (Stephenson and others 1995).

animals that have no evolutionary history of having been preyed on. In addition, it is suggested that harvesting wolves has lingering effects on the size, number, stability, and persistence of family-group social units (packs); on reproductive, hunting, and territorial behavior; on the role of learning and related traditions in wolf packs; on within-group and between-group patterns of genetic variation; and on overall mortality (Haber 1996).

Human hunting might be "unnatural" but nonetheless similar to other sources of wolf mortality. One way to evaluate the claim that hunting is more damaging to wolves than to ungulates is to compare the rates at which wolves die in hunted and unhunted populations. The demographic data on unharvested wolves presented by Haber (1996:1079-81) indicate that population turnover is high. Given an average pack size of 8, 1 litter per pack per year, a mean litter size of 5 (Boertje and Stephenson 1992), and 38% pups in packs in winter, 42% of pups and 36% of adults must die if the population is to remain stable. Where humans are not present, wolves typically die of starvation and intraspecific strife (wolves killing

TABLE 3.1   Changes in Winter Wolf Density After Various Amounts of Wolf Contro

Study Location	Precontrol Density	Minimum Postcontrol Density		
	Wolves/ 1,000 km^2	Wolves/ 1,000 km^2	Years Since control	Percentage of Precontrol Density
Areas where wolf control was carried out with quantitative precontrol data:				
South-central Alaska (GMU 13)	7.1	2.6	6	37%
East-central Alaska (GMU 20E)	8.1	4.1	1	51%
East-central Yukon	9.3	1.3	6	14%
Southern Yukon	12.4	3.6	3	29%
East-central Alaska (GMU 20A)	14.0	4.4	4	31%
Areas where extensive wolf control was carried out without quantitative precontrol data:				
South-central Alaska (GMU 13)	NA[a]	0.5	NA	NA
Western Alberta	NA	10.9	NA	NA
Kenai Peninsula, Alaska (GMU 15)	NA	0.4	NA	NA

[a]NA, Not Available.

wolves), at rates nearly as high as those during most control efforts. This suggests that much mortality resulting from control programs substitutes for mortality that would have occurred for other reasons. Haber (1996) described cases of long-term social stability among wolves in Denali National Park, such as one female who retained her alpha-female status within the pack for 13-14 years. However, based on their analysis based on genetic fingerprinting of wolf packs in and adjacent to Denali, Meier and others (1995) concluded that individual longevity and stability of sizes and composition of packs were less than previously thought, even where human disturbance is low.

Inevitably, the social organization and behavior of a species is influenced by

Density after Control Ended

Wolves/ 1,000 km^2	Years Since Control Ended	Percentage of Precontrol Density	Annual Rate of Increase, %	References
4.8	≤ 8	68%	8	Ballard and others 1987; Hicks 1994
7.4	4	91%	16	Gasaway and others 1992
10.4	5	112%	51	Farnell and others unpublished; Hayes 1995
10.9	3	88%	17	Hayes and others 1991
15.7	12	112%	11	Boertje and others 1995
6.6	12	NA	24	Burkholder 1959; Rausch 1969
23.8	4	NA	22	Bjorge and Gunson 1983, 1989
5.0	8	NA	37	Peterson and others 1984

a natural annual turnover rate of more than one-third. Haber (1996) argues that because wolves are a highly social species, harvesting them "implies a high potential for eventually reducing these complex societies to much simpler, more primitive forms, particularly when it is so likely to scramble their unusual genetic and cultural information transfer processes." He suggests that such problems will "ultimately translate into a major, long-term declining in numbers." However, no rigorous review of ecological or genetic data has indicated any long-term demographic consequences of human harvest, even when control efforts result in annual mortality higher than the average for unexploited populations (compare Haber 1994; Lehman and others 1992; Mech 1994; Meier and others 1995; Wayne and others 1995). The reason might be that local control efforts actually mimic epidemics of disease that historically must have eliminated whole subpopulations of wolves. If that is so, it should not be surprising that wolf populations can recover quickly from control efforts with no evident genetic effects.

Different methods of wolf reductions might affect social structure. For example, hunting, snaring, and trapping tend to take subordinates and dispersing individuals, whereas air-assisted wolf reductions can eliminate whole packs (McNay, ADFG, personnel communication). However, there are no published data with which to assess changes in social structure resulting from different harvest methods.

Arguments that harvest of wolves is ethically unacceptable because of the intelligence and social complexity of wolves (Haber 1996) can be neither supported nor rejected on scientific grounds. However, no available data suggest that the killing of wolves by humans has adversely affected the long-term social organization, reproductive rates, or population dynamics of the species.

Summary

The most important factor is the availability of ungulates such as moose, caribou and sheep for food. Alternative prey, such as hares and beaver can be important when ungulate populations are extremely low. The sources of wolf mortality are intraspecific strife, disease, parasites, and legal and illegal harvest by humans. Wolf populations have high reproductive rates and dispersal rates, so populations can recover rapidly from increased mortality or decreased prey availability.

Bear Ecology

Although bears are omnivores, they are members of the mammalian order Carnivora. Black bears are relatively docile and are almost always closely associated with forests, where cubs and even adults can climb trees to escape from danger. They have inhabited almost all forests of North America. Brown, or grizzly, bears are larger and generally more aggressive than black bears, a trait essential for defending their offspring or food in the more-treeless areas they inhabit.

The distribution of both species has decreased dramatically since Europeans began settling North America about 400 years ago. When pioneers settled the West, bears were viewed as a threat to be sought out and destroyed (Brown 1985; Storer and Tevis 1955). Wherever people and their livestock settled, bears were either eliminated or greatly reduced. In the lower 48 states, brown bears are found in only about 2% of their former range (Servheen 1990); and they have been extirpated in large portions of southern Canada (Banci and others 1994). Thus, the conservation of brown bears in regions inhabited by people is a major management challenge. Black bears remain in many portions of the West, but they were extirpated over much of the eastern United States and parts of Canada. In portions of the eastern states and even areas of the West, maintaining black bear populations is an increasing conservation issue. In more-remote northern and mountainous areas, where bear populations are more secure, the management of garbage and other attractants, human settlement, habitat, acceptable hunting limits, and bear effects on ungulates are issues that face wildlife managers.

Distribution and Density

The 3 most important variables in determining bear distribution and numbers appear to be habitat quality, human density, and human behavior (McLellan in press). Habitat quality is closely related to the abundance, quality, and spatial and temporal distribution of food. Because of the complexity of bears' diets, quantifying habitat quality on the basis of the availability of food is difficult. Body size has been used as a surrogate measure of habitat quality (Stringham 1980, 1990; Wielgus 1993), but it is also influenced by bear population density, genetic constraints, and seasonal availability of high-protein foods in otherwise-poor habitats.

Human density is relatively easy to measure; human behavior is much more difficult. In such areas as national parks, where human densities can be high but their behavior is closely controlled, people have relatively little effect on the numbers and distribution of bears. At the other extreme, a small number of armed people that practice poor garbage management can have serious effects on local bear populations; bears that are attracted by the garbage will be particularly vulnerable to human-caused mortality. Between those extremes lie a multitude of combinations of human densities and behaviors that influence bear numbers.

The density of brown and black bears varies widely across North America and in Alaska. The highest density of brown bears recorded was 551 bears/1,000 km^2, in Katmai National Park at the base of the Alaska Peninsula, where the abundance of salmon combined with good vegetation make optimal habitat conditions for bears (Miller and others 1997). Because Katmai is a national park, there is no legal hunting, and human visitors are strictly controlled. The other extreme of brown bear density is found along the fringe of brown bear distribution, where people are numerous and their behavior is anything but conducive to maintaining even a few bears. In other areas, low brown bear densities are due more to poor habitat than to the abundance and behavior of people. Such areas include the Tuktoyaktuk Peninsula in the Northwest Territories (Nagy and others 1983a) and the Arctic National Wildlife Refuge in northeastern Alaska, where there are only 4 bears/1,000 km^2 (figure 3.4, Miller and others 1997). Brown bear densities in Alaska and elsewhere are greatly influenced by their access to salmon. Coastal populations that feed on salmon have densities of 191-551 bears/1,000 km^2 (Miller and others 1997). Interior populations without salmon usually have

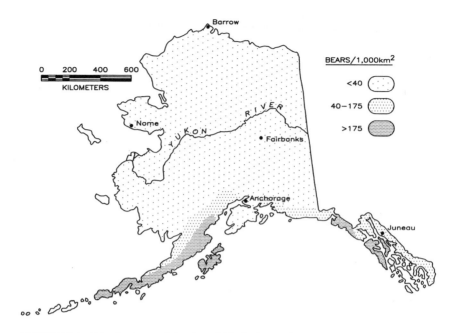

FIGURE 3.4   Distribution and density of Alaska brown bears.

densities of 4-47 bears/1,000 km^2 (McLellan 1994, Miller and others 1997). However, southern interior areas with an exceptional diversity of bear foods (McLellan and Hovey 1994) had an estimated density of 80 bears/1,000 km^2 (McLellan 1989).

Black bears are often found at much-higher densities than brown bears. On a small coastal, unhunted island of Washington state, the density of black bears (excluding cubs) reached 1,760/1,000 km^2 when the largely clearcut logged habitat was in optimal condition (Garshelis 1994; Lindzey and others 1986). On the mainland, density (excluding cubs) was estimated at 860 bears/1,000 km^2 in the Shenandoah National Park, a park with excellent bear habitat and where bears are protected (Carney 1985). The lowest recorded estimates of black bear densities have been 90/1,000 km^2, in the Susitna drainage of Alaska and in White Rock, Arkansas (Clark 1991; Miller and others 1997). The Susitna drainage is both poor black bear habitat, and has a substantial brown bear population that might restrict the black bears to a finger of streamside spruce forests (Miller and others 1997). The Arkansas study area supports a reintroduced population in which many females are shot by hunters.

Brown, or grizzly, bears are distributed over most of Alaska; exceptions are Kupreanof, Prince of Wales, and associated islands, Aleutian Islands (except Uniniak), Nunivak, and St. Lawrence Island. Except for Admiralty, Baranof,

Chichagof, and associated islands, black bears are found in most forested portions of Alaska (figure 3.4). Because they are found at low densities, move over large areas, sometimes live in densely forested areas, and hibernate in the winter, when tracks would otherwise be left, they are difficult to count. However, the Alaska Department of Fish and Game biologists have developed a successful procedure for estimating black and brown bear density in open or sparsely timbered areas. They establish a census area that is ecologically representative of the area of interest and then estimate density with a capture-mark-resight (CMR) method. Bears that are captured and radio-collared at least a year before the census constitute the marked sample. Several independent visual searches of the entire census area are conducted from a fixed-wing aircraft. On each search, the number of independent bears observed, the number of these that are radio-collared, and the total number of collared animals in the census area are recorded. Density estimates have been calculated by summing results of each search and correcting for the number of days searching (Miller and others 1987; Seber 1982). More recently, Miller and others (1997) have used statistical advances (maximal-likelihood algorithms for combining research results), which adjust estimates to account for emigration and immigration (White 1983). CMR estimates have been obtained for 15 brown bear and 3 black bear populations in 17 areas of Alaska (Miller and others 1987). Because a substantial proportion of the population must be marked in areas with low densities, the census areas are smaller than those for other large mammals on which it is less difficult to observe a significant portion of the population.

Habitat and Diet

Black and brown bears survive during the winter by retiring to dens, reducing their body temperature, and living off accumulated fat. They must ingest sufficient nutrients to meet the demands of growth and reproduction, and they must accumulate enough fat stores in a half-year to provide them with energy for the other half. Thus, during the active part of their year, feeding is their dominant activity.

Although bears have a carnivore's digestive anatomy and thus are relatively poor at digesting plant material, black and brown bears eat both plant and animal matter. The nutritional value of animal matter is higher and more consistent than plant matter, but plants are much more abundant and easier to obtain. When foraging for food, bears make tradeoffs between feeding on readily available but low-quality plant foods and seeking rare but high-quality animal matter. The quality and availability of plant food and the availability of animal matter vary widely among areas and over the seasons, and bear foraging behavior is consequently complex and dynamic.

In spring, the dominant plant foods of bears are new shoots of green vegetation, such as grasses, sedges, horsetails, and broad-leaved plants. These plants are relatively abundant but contain only about one-fifth the digestible energy per unit of dry weight as ungulates (McLellan and Hovey 1994; Pritchard and Robbins 1990). Spring is also when ungulates are born, and both black and brown bears are relatively efficient at catching newborn ungulates during their first few weeks of life (Ballard and others 1990; Boertje and others 1988; Franzmann and Schwartz 1986; Franzmann and others 1980; Gunther and Renkin 1990). In some areas, adult ungulates are also more easily captured during spring than later in summer or fall (Boertje and others 1988). In portions of coastal British Columbia and Alaska that have abundant spring plant foods but relatively few ungulates, bears expend little effort seeking calves and tend to forage more at estuaries, riparian areas, and avalanche paths (Hamilton 1988; Schoen and Beier 1990). Where both ungulates and spring plant foods are abundant, bears might forage on vegetation and, once satiated, search for ungulate calves and vulnerable adults. When spring plant foods are not readily available, bears expend more effort to find ungulate calves and adults.

As the summer progresses and ungulate calves mature, they become quicker and more difficult for bears to catch (Gunther and Renkin 1990). Concurrently, the quality of plant foods increases greatly in many areas. Over most of western North America, fruits of huckleberries (*Vaccinium* spp.), soap berries (*Shepherdia canadensis*), and a great variety of other species become ripe. Fruits, although low in protein, are rich in energy and are highly digestible (McLellan and Hovey 1994; Pritchard and Robbins 1990). In some more southern and eastern areas, seeds from white bark pine (*Pinus albicaulis*) and oaks (*Quercus* spp.) are a rich source of energy. During the summer and fall, when bears acquire the bulk of the fat needed for winter dormancy, they feed mostly on salmon in coastal areas (Jonkel and Cowan 1971; Kingsley and others 1988; Reiner 1996). Feeding on ungulates can increase in the fall during the rutting season (Boertje and others 1988; Hamer and Herrero 1991) or when hunters wound their quarry or leave gut piles (McLellan and Hovey 1994).

Social Structure and Movements

Male bears spend most of their life as solitary individuals; females are either alone or with their latest litter of cubs. Exceptions are during mating and at concentrated food sources, such as salmon spawning streams. Individuals have home ranges or areas in which they live year after year. Within their home range, bears gain knowledge of the location and seasonality of resources, such as food. The size of a home range depends largely on the sex and species of the bear and the productivity of the habitat. Brown bears generally have much larger ranges than black bears, and male bears have much larger ranges than females. Subadult male bears range over large areas when they disperse and seek permanent home ranges.

Home ranges are typically smaller in high-quality than in poorer-quality habitats. For example, average male and female brown bear home ranges were 100 and 37 km^2, respectively, on the productive coastal Admiralty Island (Schoen and Beier 1990) but 3,757 and 884 km^2 respectively in Yellowstone National Park (Blanchard and Knight 1991). In most cases, the home ranges of individuals of both sexes overlap extensively. In some areas, some female black bears maintain exclusive home ranges or territories (Rogers 1987).

Reproduction

It is well established that small species tend to reproduce much faster than large species (Charnov 1982), but bears have even lower reproductive rates than their body size would predict. Bears have long interbirth intervals and are slow to mature. Female brown bears occasionally produce their first litter when they are 4 years old, but usually they are 5-8 years old. Female black bears can have their first litter when they are 3 years old, but more commonly not until when they are 4-6 years old. In the Susitna drainage of Alaska and the Flathead drainage of British Columbia, where research has been conducted on sympatric black and brown bear populations, the ages at first litter are similar in the two species (Garshelis 1994; Hovey and McLellan unpublished data; McLellan 1994).

An average interbirth interval of 2.6 years was recorded for brown bears on the East Front of Montana, where some extremely productive females had 2-year intervals. However, average intervals of 3-4 years are more common, and average intervals of more than 4 years are recorded in some areas of Alaska, including Kodiak Island and Black Lake on the Alaska Peninsula (Aune and Kasworm 1989; McLellan 1994). Black bear interbirth intervals are about a year shorter than those of brown bears; study averages range from 2 to 3 years.

Average interbirth intervals are probably longer than have been reported. Because of the duration of research projects and the battery life of radio collars, intervals of 2 or 3 years are more commonly recorded than longer intervals (McLellan 1994; Schoen and Beier 1990). That bias is particularly evident in some brown bear studies, in which females tracked for 9 years failed to produce a second litter (Sellers 1994).

Brown bear litter sizes range from an average of 1.7 cubs/litter (Glacier National Park Montana) to 2.5 cubs/litter (Kodiak Island and Black Lake), but litters of 4 are recorded (Martinka 1974; McLellan 1989, 1994; Pearson 1975). Average black bear litter size varied greatly among study areas across North America. In Arkansas, an average of 1.4 was recorded; in Pennsylvania, the average was 3.0 (Alt 1989; Clark 1991). In summary, brown bears usually do not produce a litter until they are 5-7 years old and produce only 0.45-0.85 cub/year, and black bears reach maturity about 1 year earlier and produce 0.55 to more than 1 cub/year.

Survival Rates and Causes of Mortality

The low reproductive rate of bears is balanced by their high annual adult survival rate of more than 92% (Eberhardt and others 1994; Hovey and McLellan 1996; McLellan 1989; Sellers 1994; Sellers and others 1993; Titus and Beier 1994). Survival rates of cubs are much more variable among study areas. Annual survival of brown bear cubs range from just over 35% in Katmai National Park (Sellers and others 1993) to 87% in Montana and southern British Columbia (Aune and others 1986, Hovey and McLellan 1996). Cub survival might vary with population density, but such relationships are unclear. Brown bear densities are high in both Katmai National Park and the Admiralty Islands, but cub survival rates are 35% and 80%, respectively (Schoen and Beier 1990). Reported annual survival of black bear cubs has ranged from as low as 30% in some study areas to 90% in others (Garshelis 1994).

Bears die of a variety of causes. Natural causes include predation by other bears and other carnivores, accidents, and old age. Starvation of adult bears is rare but appears to be relatively common for young bears after years of poor food production (Schwartz and Franzmann 1991). In almost all study areas, people are the primary source of mortality for bears, legally or otherwise.

Dispersal

Dispersal rates and distances of bears are poorly known. In a Minnesota study, all male black bears dispersed 33-324 km from the study area when they were 2-4 years old (Rogers 1987). Similarly, all marked male black bears dispersed from a study area in Idaho (Beecham 1983), and male dispersal was common in Washington (Lindzey and Meslow 1977) and Arizona (LeCount 1982). In Arkansas, male black bears did not disperse from the study area (Smith and Pelton 1990). Female black bears typically remain in the maternal home range (LeCount 1982).

On the Tuktoyaktuk Peninsula, 3 subadult male brown bears were shot 25, 31, and 92 km from their mothers' home ranges (Nagy and others 1983a). In the Alaska Range, 1 subadult male brown bear was shot 32 km and a second shed his collar 37 km from maternal ranges (Reynolds and Boudreau 1990). Four subadult males moved an average of 70 km from their maternal ranges, but 1 was shot only 15 km from his maternal range (Blanshard and Knight 1991). As is apparent with black bears, subadult female brown bears tend to remain in or overlap their mothers' home range (Blanshard and Knight 1991; Nagy and others 1983b), but some females dispersed from a high-density population in the Flathead Valley (McLellan and Hovey unpublished data).

Rates of Population Change

The maximal reported rate of annual increase in a brown bear population is 8.5% in the Flathead drainage of British Columbia, an area with a great diversity of bear foods (Hovey and McLellan 1996; McLellan 1989; McLellan and Hovey 1994); this population is apparently increasing after overharvesting. Annual rates of change of 4.6% and 1.0% have been reported in Yellowstone National Park and the South Fork of the Flathead in Montana, respectively (Eberhardt and others 1994; Servheen and others 1994). Black bears reproduce at higher rates than brown bears and thus have a higher potential rate of increase; however, because of human harvest, this rate is rarely realized.

Given the great variation in brown bear reproductive rates in Alaska, there is potential for great variation in rates of increase or sustainable harvest. Population decreases due to intentional overharvest in areas with road access to hunters have been reported, as has a lack of reproduction in a poorly accessed area (F. Miller 1990; SD Miller 1995; Reynolds 1994).

Consequences of Control on Bear Populations

The results of bear control by lengthening of hunting seasons, translocations, or other methods are difficult to determine because bears are difficult to count. In addition, bear population densities are highly variable. Bears are easily overexploited and, unlike wolves, recover slowly from low densities. Coastal bears that have access to salmon have higher reproductive rates and can sustain higher harvest rates than interior bears, which lack this food source. It is the interior bears, however, that feed on ungulate calves. The low densities and low reproductive rates of these bears make bear control a delicate management decision. It is well established that bear populations can be reduced and even eradicated by overharvest. In addition, reduced reproduction due to social disruption by young immigrant male bears has been suggested in areas where brown bear numbers were reduced (Wielgus 1993; Wielgus and Bunnell 1994). It is clear that if bear control is conducted, it must be in conjunction with detailed monitoring of the bear populations to be controlled, to reduce the risk of local extirpation.

If given the opportunity, most (or all) bears would kill and eat an ungulate calf, but individual bears vary widely in predation success. Boertje and others (1988) found that 4 of the 9 bears they tracked killed 72% of moose calves that were killed by the 9. Ballard and others (1990) found wide variation in individual bear's predation rates. Many black and brown bears killed no moose calves while they were monitored, whereas 1 female brown bear killed 1 moose calf every 2.8 days and a black bear killed 1 moose calf every 5.7 days. Ballard and Miller (1990) noted that 2 identifiable brown bears were known to have killed 6 of the 12 radio-collared moose calves that died in their study. In a foraging study conducted in Yellowstone National Park, French and French (1990) observed that some brown bears never fed on elk, whereas others fed almost exclusively on elk. Given this variation among bears, the outcome of bear control programs is highly unpredictable. It will depend on which bears are removed and the feeding habits of the removed bears. Not surprisingly, bear control experiments have yielded inconsistent and short-term effects on ungulate calf survival (Ballard and Miller 1990; Crête and Jolicoeur 1987; Miller and Ballard 1992; Schlegel 1983).

Summary

The diet of bears includes neonate and adult ungulates, especially moose and caribou. Salmon are important when they are available, as well as ground squirrels and other small mammals. Bears also eat grasses, sedges, broad-leaved plants, fruits and seeds. Sources of mortality besides legal and illegal hunting by humans include intraspecific strife among bears, accidents, and rarely predation by wolves and disease. Humans affect the distribution of bears by their hunting pressure and how they manage garbage.

Caribou Ecology

Caribou (*Rangifer tarandus*) inhabit the high Arctic, tundra, boreal and subboreal forests, and wet interior mountains across the Northern Hemisphere. Caribou are found in most regions of Alaska (figure 3.5, table 3.2). All North American caribou are wild, except in a few areas where semi-domesticated varieties called reindeer have been introduced for herding by native peoples. In Eurasia, there are large herds of reindeer (Andreyev 1977). During the 1980s, most caribou populations in Alaska were increasing (Davis and Valkenburg 1991). In the early 1990s, most of the interior Alaskan herds began declining, while many of the smaller herds and the Western Arctic and Mulchatna herds (two of the largest herds in Alaska) continued to increase (Valkenburg and others 1996, table 3.2).

Social Structure and Movements

Caribou generally exist in separate herds or populations that annually move over extensive areas, sometimes migrating hundreds of kilometers between wintering areas and calving-summering grounds, behavior that is probably an antipredator strategy (Heard and others 1996; Seip and Brown 1996). Female caribou usually show strong fidelity to the specific calving grounds of the herd, returning each year to the same general area to give birth (Skoog 1968). Some herds migrate between summer and winter ranges, and all caribou shift their ranges in response to forage conditions. Short-term variation in habitat use and movements also can be stimulated by harassment by wolves (Scotter 1995).

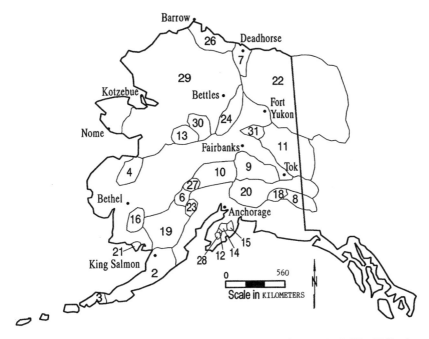

FIGURE 3.5   Approximate ranges of caribou herds in Alaska. SOURCE: Valkenburg and others 1996. Printed with permission from the author.

Caribou of the Western Arctic Caribou Herd in Alaska select calving grounds that have low snow cover and early snow melt (Lent 1980).

Knowing whether caribou regularly switch between herds is important to wildlife managers because if substantial numbers of caribou emigrate from herds that are increasing to areas where herds are small, resources used to increase the size of small herds might be wasted; immigration would "rescue" the depressed herds. However, although caribou are highly migratory and herds sometimes intermingle, direct evidence of more than the occasional individual leaving one herd to join another is lacking. Exchange of a few individuals between herds is important from a genetic standpoint, but periodic exchanges of large numbers of individuals would be necessary to alter herd population dynamics. If large shifts of animals were regular, they should have been detected from the extensive radio-collaring of caribou in the Nelchina, Delta, Ashihik, and Fortymile herds.

Nevertheless, caribou herds can undergo rapid population fluctuations. For example, the Mulchatna Herd increased from about 30,000 to 110,000 animals between 1984 and 1993, and its range shifted to the west. When the Fortymile Herd had 570,000 animals and ranged over an area 5 times as large as it occupies today (Boertje and Gardner 1996; Valkenburg and Davis 1986), it might have engulfed other herds. Maps in Valkenburg and Davis (1986) show that the

TABLE 3.2   Estimated Size and Approximate Density of Alaskan Caribou Herds

Herd No.	Herd Name	1993 Population Estimate	Population Trend Since 1989
1	Adak	750	up
2	Alaska Peninsula (North)	18,000	down
3	Alaska Peninsula (South)	2,500	stable
4	Andreafsky	<50	unknown
5	Beaver Mountains	649	unknown
6	Big River	750	unknown
7	Central Arctic	23,444	stable
8	Chisana	850	down
9	Delta	3,661	down
10	Denali	1,890	down
11	Fortymile	21,884	down
12	Fox River	75	up
13	Galena Mountain	275	up
14	Kenai Lowlands	100	stable
15	Kenai Mountains	300	stable
16	Kilbuck Mountains	2,500	up
17	Macomb	500	down
18	Mentasta	880	down
19	Mulchatna	110,000	up
20	Nelchina	40,361	stable
21	Nushugak Peninsula	750	up
22	Porcupine	165,000	stable
23	Rainy Pass	500-1,000	unknown
24	Ray Mountains	700	up
25	Sunshine Mountains	800	unknown
26	Teshepuk	27,630	up
27	Tonzona	800	down
28	Killey River	100	up
29	Western Arctic	450,000	up
30	Wolf Mountain	650	stable
31	White Mountains	1,000	up
	TOTAL (approximate)	880,000	

SOURCE: Valkenburg and others 1996.

Fortymile Herd historically occupied much of the current wintering area of the Porcupine Herd in the Ogilvie Mountains. Heard and Calef (1986) suggested, in the absence of reliable census and management data, that the increase in the Kaminuriak Herd on the calving grounds from 13,000 in 1980 to 41,000 in 1982 must have included immigration from other herds. Continued extensive monitoring of radio-collared animals and genetic analyses will assist in determining the frequency and extent of exchanges of individual animals between herds. Meanwhile, it appears imprudent to rely on immigration as a substantial source of growth of depressed herds during time frames of a few years.

Habitat and Diet

Caribou are primarily grazers, but they occasionally browse. They eat a variety of species—lichens, willows, forbs, sedges, grasses, mosses, blueberries, and mushrooms (Klein 1970; White and Truddell 1980). In summer, caribou forage primarily in areas with high species diversity and in microhabitats that provide high-quality forage (White and Trudell 1980). In the winter, they move to areas dominated by lichens (Boertje 1984; Holleman and Luick 1977; Klein 1982). In southern boreal forests or in subalpine habitats in Canada, arboreal lichens often make up the bulk of the caribou winter diet; in tundra areas and in the northern fringes of the boreal forest, terrestrial lichens dominate (Boertje 1984; Klein 1982). Lichens can be adversely affected by overgrazing and trampling (Pegau 1970, 1975). Stands of lichens have deteriorated from excessive grazing by caribou and trampling by caribou, moose, and domesticated reindeer (Gaare and Skogland 1980; Pegau 1975; White and Trudell 1980).

Caribou have especially high nutritional requirements during the summer because their offspring grow very rapidly (McEwan 1968). Consequently, in summer, caribou are selective feeders, seeking out new leaves of willow and sedges and buds and flowers of the sedge known as cotton grass and other flowering plants of the tundra (Klein 1970). Caribou move to southern exposures at the beginning of the season, where plant growth begins sooner. By late summer, nutritional needs of caribou have shifted from a high protein requirement for milk production and body and antler growth to an emphasis on carbohydrates needed to lay down fat stores for the coming winter.

In winter, caribou dig through snow to obtain food; because they are constantly on the move in very cold weather, they require forage of high energy value (Klein 1992). Lichens are an ideal winter food for caribou because they are high in carbohydrates that provide the energy to meet their metabolic needs in winter. Although lichens are poorly digested and avoided by most other herbivores, caribou are well adapted to digest them and have unique microorganisms to aid the process in their rumens.

Caribou can maintain their body temperatures at ambient temperatures of at least −35°C without increasing their metabolic rates (Hart and others 1961). Moose are less well insulated, but they can maintain their body temperatures without increasing metabolic rates when ambient temperature is as low as −32°C (Renecker and Hudson 1986). At still lower air temperatures, they must expend extra energy to stay warm, and strong winds effectively lower ambient temperatures (Bergerud 1983).

During summer, caribou are attacked by mosquitoes, warble and bot flies, and biting flies, particularly when weather conditions are favorable for insect flight. Caribou respond to mild harassment by mosquitoes by feeding while facing the wind (White and Milan 1981), but when insect harassment is severe, caribou stop feeding and seek relief by running, standing in water or on snow patches, forming tight herds (with constant jostling), or moving to windy areas (Batzli and others 1981; Helle and others 1992; White and others 1975). Those activities interfere with growth and fattening (Reimers 1980) during summer, when energy requirements are highest (Boertje 1985). Body fat reserves of caribou in the Northwest Territories decline to almost zero in July; when insect numbers decline in August caribou reach their highest yearly levels of body fat (Heard and others 1996).

The effect of snow on wildlife has been studied extensively in Europe and North America (Peek 1974; Pruitt 1981). Snow accumulations vary with climate, storm tracks, moisture content, redistribution by wind, topography, and forest cover. Snow adversely affects moose and caribou by burying plants, by reducing the animals' ability to locate plants by smell (Leader-Williams and others 1981), and by increasing the energy cost of locomotion (Parker and Robbins 1984; Telfer and Kelsall 1984). Deep snow that has a hard-packed surface can increase ungulate vulnerability to wolf predation because wolves, but not moose and caribou, can run over the surface without sinking (Mech and others 1995; Nelson and Mech 1986).

Caribou populations are often regulated by winter food supplies. In addition, reproductive rates are lower and juvenile survival rates are higher when summer food supplies are low. Caribou are smaller where they live at high densities (Skogland 1985).

Summary

In summer caribou eat broad-leaved plants, willows and sedges and their supply and quality may vary in relation to both previous range conditions and the number of shady days in the current growing season. Nutrient quality of some forage plants is improved by more shady days. The winter diet is largely slow-growing lichens in areas that have not been burned for more than 50 years. Caribou are vulnerable to starvation and predation by wolves in severe winters and to predation on neonates and adults on calving grounds. Brown bears and black bears also take neonate caribou, as do eagles. Caribou numbers can also be affected by outbreaks of mosquitoes, flies and disease, and by other human disturbances associated with roads and oil fields.

Moose Ecology

Distribution and Density

Moose occur throughout northern mountains, boreal forest, and riparian areas of the Northern Hemisphere. Moose appear to have been absent or rare in many areas of Alaska as recently as 75 years ago (Huntington and Rearden 1993, Spalding 1990). Today moose are most abundant on the Kenai Peninsula, the lower Matanuska and Susitna drainages, and locally throughout the interior and western Alaska in riparian communities associated with major river systems.

Social Structure and Movements

Most of the year moose are solitary or live in small groups of bulls or of females and offspring (Miquelle and others 1992). The breeding season of moose varies with locality, but generally begins in late September and continues to early November (Van Ballenberghe and Miquelle 1993). During this period, mature bulls devote most of their time to seeking copulation and can lose up to 25% of their body mass. Females continue to feed and gain body condition that will be needed for pregnancy over the lean winter months. During the rut, female moose in open habitats may be aggregated into larger groups of breeding harems by dominant bulls. In winter, moose sometimes assemble into larger herds. Moose are typically not migratory, although they can shift between summer and winter ranges. Some moose that summer on alpine ranges winter at lower elevations.

Habitat and Diet

Moose are most abundant in early successional-stage habitats that develop after fires, floods, logging, and land-clearing. Moose eat primarily leaves and tender stems of woody plants, but during summer they also eat herbaceous plants. In winter, they prefer to browse on species that are relatively high in protein and carbohydrates, such as willow, aspen, and birch; the former are found in thickets along rivers. During late summer, moose accumulate fat deposits for use in winter and, among adult males, for use in the rut. The winter diet of moose is dominated by twigs of woody shrubs grown during the previous summer. The bark and buds of these young woody twigs contain high concentrations of available nutrients.

Summary

Moose depend on access to shrubby plant growth, especially willows, whose availability is best 10-30 years after a fire, but is also affected by browsing history by moose, snowshoe hares and the recent history of snow depth and condition. Plant nutrient quality is improved if there have been more shady days in the growing season (Molvar and others 1993). During plant succession conifers crowd out deciduous browse. Improvement of resources for moose must be considered in their management.

Long winters and deep crusted snow can increase mortality or weaken moose so that they are more vulnerable to predation. Levels of predation by wolves are affected in turn by human harvest of wolves and other factors that affect wolf populations. If moose have access to wetland habitats, they can more readily escape from predation by wolves. Brown and black bears often prey heavily on moose, especially newly born calves in some areas.

Comparative Ecology of Moose and Caribou

Reproduction, Survival Rates, and Causes of Mortality

Moose and caribou differ strikingly in their basic reproductive biology (table 3.3). In habitats of moderate to high quality, many female moose reach breeding age in their second year and give birth to their first calf at the age of 2 years (Franzmann 1981). In later pregnancies, twinning is common and triplets are occasionally born. Female moose with more than 1 calf, however, are more likely to lose a calf to predators, accidents, or other causes than are females with a single calf.

Caribou, in contrast, always give birth to a single calf, and most females do

TABLE 3.3   Comparative Population Characteristics of Moose and Caribou[a]

Characteristic	Moose	Caribou
Age at first breeding	1 yr 4 mo	2 yr 5 mo
Gestation duration	8 mo	7.5 mo
Age at birth of first young	2 yr	3 yr
Frequency of births	Annual	Annual
Litter size	1-3	1
Twinning rates, %	17-63	0
Longevity, yr:		
In captivity or without predation:	20+	21+
In natural habitats:	male,7; female 10	male,7; female,10
Productivity, calves/100 cows	60-90 (2+ yr)	50-90 (3+ yr)
Survival rates, mean %:		
Calves	32	20
Adults	90	90

[a]Data from Ballard and others 1991; Bergerud 1980; Boertje and others 1995; Dauphine 1976; Franzmann 1981.

not reach breeding age until their third year, although under favorable nutritional conditions some female caribou bear their first calves at the age of 2 years. In both caribou and moose, the pregnancy rate increases until females reach the age of 4 or 5 years and declines in very old animals (Dauphine 1976; Franzmann 1981). Because of those differences, moose populations can increase and recover from depression faster than caribou if the quality and quantity of forage are adequate.

The ratio of calves to cows is relatively easy to obtain, but identifying the cause of a low ratio is difficult and expensive. Calves that are born dead are more easily discovered than calves taken by predators. Nevertheless, predation rates are known to be high among calves during their first few weeks of life (tables 3.4 and 3.5). Compared to wolves, predation by bears is a significantly greater source of mortality for moose calves and, in some cases, for caribou calves (tables 3.4 and 3.5; Osborne and others 1996).

The high annual survival rates of adult female moose and caribou, which are usually above 85%, mean that these species can thrive even when calf mortality is substantial, because most cows will live to reproduce another year, but excessive calf mortality can cause a population collapse (Bergerud 1996; Larsen and others 1989).

TABLE 3.4 Cause-Specific Mortality (%) in Radio-Collared Caribou Calves in Areas of Southern Alaska and British Columbia

| Area | No. of Calves | Predator | | | | Other Mortality | | TOTAL (%) |
		Wolf (%)	Brown Bear (%)	Other (%)	Un-known (%)	Not Preda-tion (%)	Un-known Cause (%)	
Denali[a]	226	14	20	2	9	1	4	50
Delta[b]	93	25	22	15	0	0	4	66
Fortymile[c]	109	24	17	12	0	6	6	65
Spatsizi[d]	34	24	9	0	0	15	9	57
Average		22.0	17.0	7.3	2.3	5.5	5.8	59.5

[a]Adams and others (1995): birth to 1 yr of age.
[b]Valkenburg and others (1996): birth to September 30.
[c]Boertje and Gardner (1996): birth to 1 yr of age.
[d]Page (1985): birth to 4 months of age.

TABLE 3.5 Cause-Specific Mortality (%) in Radio-Collared Moose Calves in Alaska and the Yukon

| Area | No. of Calves | Predator | | | | | Other Mortality | | TOTAL |
		Wolf (%)	Brown Bear (%)	Black Bear (%)	Other (%)	Unknown (%)	Not Predation (%)	Unknown (%)	(%)
Susitna, June-Oct.[a]	198	3	44	2	1	2	8	3	63
Susitna, *Nov.-May*	114	0	2	0	0	1	10	0	13
Yukon[b]	117	18	42	3	0	3	6	9	81
Tanana[c]	40	0	10	13	0	0	0	10	33
GMU20E[d]	33	12	52	0	0	0	12	0	79
Kenai, 1947[e]	47	6	6	34	0	2	4	4	69
Kenai, 1969[f]	74	1	3	35	0	5	5	1	50
Average[g]		6.7	26.2	15.0	0.2	2.0	5.8	4.5	60.3

[a]Ballard and others (1991): wolves had been decreased, bears also decreased late in neonatal mortality study.
[b]Larsen and others (1989): annual survival rates were measured during a period when 55% of the wolves were removed annually. 29.9 calves/100 cows in fall and 19.5 calves/100 cows in spring.
[c]Keech (in preparation): calf survival was measured from birth to 4 mo of age.
[d]Gasaway 1992): wolves had been reduced by 60%; 22 calves/100 cows in winter; adult mortality of 9%; to 11 months of age.
[e]Franzmann and others (1980): 1947 burn; survival to about 2-2.5 mo of age.
[f]Franzmann and Schwartz (1986): 1969 burn; survival to 4 mo of age.
[g]Average excludes Susitna, November-May.

Habitat Quality and Ungulate Population Dynamics

Habitat quality is an important determinant of dynamics of populations of large mammalian herbivores (Caughley 1981; Crête and others 1990; Jeffries and others 1994; Pastor 1988). Knowledge of seasonal and annual variability in forage quality and quantity is important for development of management plans for ungulate populations. In Alaska, habitat carrying capacities have been assessed primarily by such indirect indicators as the length of the lower jaw, marrow fat, and body weight of individual animals, or such demographic data as birth rate, calf survival, or population age structure. Those indirect measures, which assume that the status of the animals reflects the status of their habitats, have proved useful when time, expertise, or funding was not available for direct assessment of the habitats. However, those measures are often slow to respond to habitat changes, can be unduly influenced by short-term weather anomalies, and are influenced by factors other than habitat quality.

A few long-term studies involving experimental manipulations and monitoring of habitat changes caused by these and natural disturbances, such as fire, have been undertaken. The collaborative research on moose-habitat relationships by ADFG and the Fish and Wildlife Service (FWS) in the Kenai National Wildlife Refuge (formerly the Kenai National Moose Range) is an excellent example. ADFG, with management responsibility for wildlife, and FWS, with responsibility for habitat, entered into a cooperative agreement that yielded basic information on responses of moose habitat to fire and other manipulations in the Kenai Peninsula and on the physiological, nutritional, and population consequences of habitat changes, constraints, and stocking levels (Oldemeyer and Regelin 1987; Regelin and others 1987; Schwartz and others 1988). Information gained from this long-term effort has been extremely valuable in the management of moose in Alaska and especially on the Kenai Peninsula.

Intensive habitat-relationship studies like those carried out on the Kenai need not be duplicated in each area where ADFG plans to intensify management activities, but some level of knowledge of the existing habitat is essential if wise decisions are to be made. Desirable information includes description of vegetation, fire history, proportions of habitat in different successional stages after fire, rough biomass estimates of critical forage types (for example, lichens on caribou winter ranges and primary winter browse species in moose habitats), and distribution and depth of snow cover in the winter.

REFERENCES

Adams LG, BW Dale, and LD Mech. 1995. Wolf predation on caribou calves in Denali National Park, Alaska. *In* LN Carbyn, SH Fritts, and DR Seip, Eds. Ecology and Conservation of wolves in a changing world: proceedings of the second North American symposium on wolves. Canadian Circumpolar Inst, Univ Alberta, Edmonton, 1995.

Alaska Bureau of Vital Statistics. 1995. Population trends in Alaska.

Alt GL. 1989. Reproductive biology of female black bears and early growth and development of cubs in Northeastern Pennsylvania. Ph.D. thesis, WV Univ, Morgantown. 116 Pp.

Andreyev VN. 1977. Reindeer pastures in the subarctic territories of the USSR. Pp. 275-313 *in* W Krause, Ed. Handbook of vegetation science, part XIII. Application of vegetation science to grassland husbandry. Dr. W. Junk BV publishers, The Hague, Netherlands.

Aune K and W Kasworm. 1989. Final report. East front grizzly studies. Mont Dep Wildl Parks, Helena. 332 Pp.

Aune K, M Madel, and C Hunt. 1986. Rocky mountain front grizzly bear monitoring and investigation. Mont Dep Wildl Parks, Helena. 175 Pp.

Ballard WB, JS Whitman, and CL Gardner. 1987. Ecology of an exploited wolf population in south-central Alaska. Wildlife Monographs 98:1-54.

Ballard WB and SL Miller. 1990. Effects of reducing brown bear density on moose calf survival in south-central Alaska. Alces 26:9-13.

Ballard WB, SD Miller, and JS Whitman. 1990. Brown and black bear predation on moose in southcentral Alaska. Alces 26:1-8.

Ballard WB, JS Whitman, and DJ Reed. 1991. Population dynamics of moose in southcentral Alaska. Wildlife Monographs 114:1-49.

Ballard WB, CL Gardner, and DJ Reed. 1995. Use of line-intercept transects for estimating wolf densities. *In* LN Carbyn, SH Fritts, and DR Seip, Eds. Ecology and conservation of wolves in a changing world. Canadian Circumpolar Inst., Univ. Alberta, Edmonton.

Banci V, DA Demarchi, and WR Archibald. 1994. Evaluation of the population status of grizzly bears in Canada. Int Conf Bear Res and Management 9:129-142.

Batzli GO, RG White, and FL Bunnell. 1981. Herbivory: a strategy of tundra consumers. International Biological Programme 25:359-375.

Becker E and others. In press. A population estimator based on network sampling of tracks in the snow. J Wildlife Management.

Beecham JJ. 1983. Population characteristics of black bears in west central Idaho. J Wildlife Management 47:405-412.

Berg WE and DW Kuehn. 1982. Ecology of wolves in northcentral Minnesota. *In* FH Harrington and PC Paquet, Eds. Wolves: a worldwide perspective of their behavior, ecology and conservation. Noyes Publications, Park Ridge, NJ.

Bergerud AT. 1974. Decline of caribou in North America following settlement. J Wildlife Management 38:757-770.

Bergerud AT. 1980. A review of the population dynamics of caribou and wild reindeer in North America. Proc Int Reindeer/Caribou Symp 2:556-581.

Bergerud AT. 1983. The natural population control of caribou. *In* FL Bunnell, DS Eastman, and JM Peek, Eds. Symposium on natural regulation of wildlife populations. Proc No. 14. Forest Management and Range Exp Stn, Univ Idaho, Moscow.

Bergerud AT. 1996. Evolving perspectives on caribou population dynamics, have we got it right yet? Rangifer Special Issue 9.

Bjorge RR and JR Gunson. 1983. Wolf predation of cattle in the Simonette River Pastures in northwestern Alberta. *In* LN Carbyn, Ed. Wolves in Canada and Alaska. Canadian Wildlife Service Report Series 45. 135 Pp.

Bjorge RR and JR Gunson. 1989. Wolf, *Canis lupus*, population characteristics and prey relationships near Simonette River, Alberta. Canadian Field-Naturalist 103:327-334

Blanchard BM and RR Knight. 1991. Movements of Yellowstone grizzly bear. Biol Cons 58:41-67.

Boertje RD. 1984. Seasonal diets of the Denali caribou herd, Alaska. Arctic 37:161-165.

Boertje RD. 1985. Seasonal activity of the Denali Caribou herd, Alaska. Rangifer 5:32-42.

Boertje RD, WC Gasaway, DV Grangaard, and DG Kelleyhouse. 1988. Predation on moose and caribou by radio-collared grizzly bears in eastcentral Alaska. Canadian J Zoology 66:2492-2499.

Boertje RD and RO Stephenson. 1992. Effects of ungulate availability on wolf reproductive potential in Alaska. Canadian J Zoology 70:2441-2443.

Boertje RD, DG Kelleyhouse, and RD Hayes. 1995. Methods for reducing natural predation on moose: an evaluation. *In* LN Carbyn, SH Fritts, and DR Seip, Eds. Ecology and conservation of wolves in a changing world. Canadian Circumpolar Inst, Univ Alberta, Edmonton.

Boertje RD and CL Gardner. 1996. Factors limiting the Fortymile caribou herd. Alaska Dep Fish and Game. Fed Aid in Wildlife Restor. Prog Rep. Proj W-24-4. Study 3.38. Juneau. 79 Pp.

Brand CJ, MJ Pybus, WB Ballard, and RO Peterson. 1995. Infectious and parasitic diseases of the gray wolf and their potential effects on wolf populations in North America. *In* LN Carbyn, SH Fritts, and DR Seip, Eds. Ecology and conservation of wolves in a changing world. Canadian Circumpolar Inst, Univ Alberta, Edmonton.

Brown DE. 1985. The grizzly in the southwest; documentary of an extinction. Univ of Okla Press, Norman. 274 Pp.

Bryant JP and PJ Kuropat. 1980. Selection of winter forage by subarctic browsing vertebrates: the role of plant chemistry. Annual Review of Ecology and Systematics 11: 261-286.

Burkholder BL. 1959. Movements and behavior of a wolf pack in Alaska. J Wildlife Management 23:1-11.

Carney DW. 1985. 1985. Population dynamics and denning ecology of black bears in Shenandoah National Park, Virginia. M.S. thesis, VA Polytech Inst and State Univ, Blacksburg. 84 Pp.

Caughley G. 1981. Overpopulation. *In* Problems in management of locally abundant wild mammals. R Jewell and S Holt, Eds. Academic Press, NY

Charnov EL. 1982. The theory of sex allocation. Monographs in population biology. Vol. 18. Princeton, NJ, Princeton University Press,

Clark. 1991. Ecology of two black bear (*Ursus americanus*) populations in the interior highlands of Arkansas. Ph.D. thesis, Univ Ark, Fayetteville. 228 Pp.

Coley PD, JP Bryant, and FS Chapin III. 1985. Resource availability and plant anti-herbivore defense. Science 230: 895-899.

Crête M and H Jolicoeur. 1987. Impact of wolf and black bear removal on cow:calf ratio and moose density in southwestern Québec. Alces 23:61-87.

Crête M and others. 1990. Food selection during early location by caribou calving in the tundra in Quebec. Arctic 43:60-65.

Dale BW, LG Adams, and RT Bowyer. 1995. Winter wolf predation in a multiple ungulate prey system, Gates of the Arctic National Park, Alaska. *In* LN Carbyn, SH Fritts, and DR Seip, Eds. Ecology and conservation of wolves in a changing world. Canadian Circumpolar Inst, Univ Alberta, Edmonton.

Dauphine TC, Jr. 1976. Biology of the Kaminuriak population of barren-ground caribou. Part 4. Growth, reproduction and energy reserves. Canadian Wildl Serv Rep. Ser No. 38. 69 Pp.

Davis JL and P Valkenburg. 1991. A review of caribou population dynamics in Alaska emphasizing limiting factors, theory, and management implications. Pp.184-209 *in* Proceedings of the Fourth North American Caribou Workshop. C Butler and SP Mahoney, Eds. Newfoundland and Labrador Wildlife Division, St. John's.

Dhondt AA. 1988. Carrying capacity: a confusing concept. Acta Oecologica 9:337-346.

Eberhardt LL, BM Blanchard, and RR Knight. 1994. Population trend of the Yellowstone grizzly bear as estimated from reproductive and survival rates. Canadian J Zoology. 72:360-363.

Edenius L. 1993. Browsing by moose on scots pine in relation to plant resource availability. Ecology 74:2261-2269.

Edenius L, K Danell, and R Bergstrom. 1993. Impact of herbivory and competition on compensatory growth in woody plants winter browsing by moose on Scots Pine. Oikos 66:286-292

Franzmann AW, CC Schwartz, and RO Peterson. 1980. Moose calf mortality in summer on the Kenai Peninsula. J Wildlife Management 44:764-768.

Franzmann AW. 1981. *Alces alces*. Mammalian Species 154:1-7.

Franzmann AW and CC Schwartz. 1986. Black bear predation on moose calves in highly productive versus marginal moose habitats on the Kenai Peninsula, Alaska. Alces 22:139-153.

French SP and MG French. 1990. Predatory behavior of grizzly bears feeding on elk calves in Yellowstone National Park, 1986-88. Int Conf Bear Res and Management 8:335-341.

Fritts SH and LD Mech. 1981. Dynamics, movements, and feeding ecology of a newly-protected wolf population in northwestern Minnesota. Wildlife Monographs 80. 79 Pp.

Fritts SH. 1983. Record dispersal by a wolf from Minnesota. J Mammalogy 64:166-167.

Fritts SH, WJ Paul, LD Mech, and DP Scott. 1992. Trends and management of wolf-livestock conflicts in Minnesota. US Department of Interior, US Fish and Wildlife Service Resource Publication 181. 27 Pp.

Fritts SH and LN Carbyn. 1995. Population viability, nature reserves, and the outlook for gray wolf conservation in North America. Restoration Ecology 3:26-38.

Fuller TK and WJ Snow. 1988. Estimating wolf densities from radiotelemetry data. Wildlife Soc Bull 16:367-370.

Fuller TK. 1989a. Population dynamics of wolves in north-central Minnesota. Wildlife Monographs 105:1-41.

Fuller TK. 1989b. Denning behavior of wolves in north-central Minnesota. American Midland Naturalist 121:184-188.

Fuller TK. 1995. Guidelines for gray wolf management in the Northern Great Lakes Region. International Wolf Center Technical Pub 271, Ely, MN. 19 Pp.

Gaare E and T Skogland. 1980. Lichen-reindeer interaction studies in a simple case model. *In* E Reimers, E Gaare, and S. Skjenneberg, Eds. Proceedings of the Second International Reindeer/Caribou Symposium, Direktoratet for vilt og ferskvannfisk, Trondheim, Norway.

Garshelis DL. 1994. Density dependent population regulation of black bears. *In* M Taylor, Ed. Density dependent population regulation in black, grizzly and polar bears. Int Conf Bear Res and Management. Monogr Series No. 3. 43 Pp.

Gasaway WC, RO Stephenson, JL Davis, PEK Shepherd, and OE Burris. 1983. Interrelationships of wolves, prey and man in interior Alaska. Wildlife Monographs 84:1-50.

Gasaway WC, RD Boertje, DV Grangaard, DG Kellyhouse, RO Stephenson, and DG Larsen. 1992. The role of predation in limiting moose at low densities in Alaska and Yukon and implications for conservation. Wildlife Monographs 120:1-59.

Gese EM, and LD Mech. 1991. Dispersal of wolves (*Canis lupus*) in northeastern Minnesota, 1969-1989. Canadian J Zoology 69:2946-2955.

Greenberg JH. 1987. Language in the Americas. Stanford University Press, Stanford, CA.

Gunther KA and RA Renkin. 1990. Grizzly bear predation on elk calves and other fauna of Yellowstone National Park. Int Conf Bear Res and Management 8:329-334.

Haber GC. 1994. Biological impacts on wolves of exploitation and control. P. 36 *in* First Annual Conference of the Wildlife Society, Albuquerque, NM (abstract).

Haber GC. 1996. Biological, conservation, and ethical implications of exploiting and controlling wolves. Conservation Biology 10:1068-1081.

Hamer D and S Herrero. 1991. Elk, *Cervus elaphus*, calves as food for grizzly bears, *Ursus arctos*, in Banff National Park, Alberta, Canada. Canadian Field-Naturalist 105:101-103.

Hamilton AN. 1988. Classification of coastal grizzly bear habitat for forestry interpretations and the role of food in habitat use by coastal grizzly bears. M.Sc. thesis, University of British Columbia.

Harrington FH, LD Mech, and SH Fritts. 1983. Pack size and wolf pup survival: their relationship under varying ecological conditions. Behavioral Ecology and Sociobiology 13:19-26.

Hart JS, O Heroux, WH Cottle, and CA Mills. 1961. The influence of climate on metabolic and thermal responses of infant caribou. Canadian J Zoology 39:845-856.

Hayes RD, A Baer, and DL Larsen. 1991. Population dynamics and prey relationships of an exploited and recovering wolf population in the southern Yukon. Yukon Fish and Wildlife Br Final Rep TR-91-1. Whitehorse. 67 Pp.

Hayes RD. 1995. Numerical and functional response of wolves, and regulation of moose in the Yukon. M.Sc. thesis, Simon Fraser University, Burnaby, British Columbia.

Heard DC and GW Calef. 1986. Population dynamics of the Kaminuriak northwest territories Canada caribou herd 1968-1985. Rangifer. Special Issue 1:159-166.

Heard DC, TM Williams, and DA Melton. 1996. The relationship between food intake and predation risk in migratory caribou and implications to caribou and wolf population dynamics. Rangifer. Special Issue 9:37-44.

Helle T, J Aspi, K Lempa, and E Taskinen. 1992. Strategies to avoid biting flies by reindeer: field experiments with silhouette traps. Annales Zoologici Fennici 29:69-74.

Hicks MV. 1994. Federal aid in wildlife restoration survey-inventory management report. Alaska Department of Fish and Game, Div Wildlife Cons, Juneau, AK.

Holleman DF and JR Luick. 1977. Lichen species preference by reindeer. Canadian J Zoology 55:14-26.

Hovey FW and BN McLellan. 1996. Estimating population growth of grizzly bears from the Flathead River drainage using computer simulations of reproduction and survival rates. Can J Zool 74:1409-1416.

Huntington S and Rearden 1993. Shadows on the Koyukuk. Alaska Northwest Books. Anchorage, AK

Jeffries R and others. 1994. Vertebrate herbivores and northern plant communities: reciprocal influences and responses. Oikos 71:193-206.

Jonkel CJ and IM Cowan. 1971. The black bear in the spruce-fir forest. Wildlife Monographs 27:1-57.

Keith LB. 1974. Some features of population dynamics in mammals. Trans Int Congr Game Bio 11:17-58.

Keith LB. 1983. Population dynamics of wolves. Pp. 66-77 *in* LN Carbyn, Ed. Wolves in Canada and Alaska, Canadian Wildlife Service Report Series 45. 135 Pp.

Kingsley MCS, JA Nagy, and HV Reynolds. 1988. Growth in length and weight of northern brown bears—Differences between sexes and populations. Canadian J Zoology 66:981-986.

Klein DR. 1968. The introduction, increase and crash of reindeer on St. Matthew Island. J Wildlife Management 32:350-367.

Klein DR. 1970. Interactions of Rangifer-Tarandus reindeer and caribou with its habitat in Alaska. Riistatteteellisia Julkaisuja. 289-293.

Klein DR. 1982. Fire, lichens, and caribou. J Range Management 35:390-395.

Klein DR. 1992. Comparative ecological and behavioral adaptations of Ovibos-Moschatus and Rangifer Tarandus. Rangifer 12:13-25.

Klein DR, DF Murray, RH Armstrong, and BA Anderson. 1997. Alaska. *In* MJ Mac, PA Opler, and PD Doran, Eds. National status and trends report. US National Biological Service, Washington, DC.

Kolenosky GB. 1972. Wolf predation on wintering deer in east-central Ontario. J Wildlife Management 36:357-369.

Krebs CJ. 1994. Ecology, the experimental analysis of distribution and abundance, 4th ed. Harper Collins College Publishers, NY, NY. 651 Pp.

Larsen DG, DA Gauthier, and RL Markel. 1989. Causes and rate of moose mortality in the southwest Yukon. J Wildlife Management 53:548-557.

Leader-Williams N, TA Scott, and RM Pratt. 1981. Forage selection by introduced reindeer on South Georgia, South Atlantic Ocean and its consequences for the flora. J Applied Ecology 18:83-106.

LeCount AL. 1982. Characteristics of a central Arizona black bear population. J Wildlife Management 46:861-868.

Lehman NE, P Clarkson, LD Mech, TJ Meier, and RK Wayne. 1992. A study of the genetic relationships within and among wolf packs using DNA fingerprinting and mitochondrial DNA. Behavioral Ecology and Sociobiology 30:83-94.

Lent PC. 1980. Synoptic snowmelt patterns in Arctic Alaska in relation to caribou habitat use. Pp. 71-77 in E Reimers, E Gaare, and S. Skjenneberg, Eds. Proceedings of the Second International Reindeer/Caribou Symposium, Direktoratet for vilt og ferskvannfisk, Trondheim, Norway.

Licht DS and SH Fritts. 1994. Gray wolf (Canis lupus) occurrences in the Dakotas. American Midland Naturalist 132:74-81.

Lindzey FG, KR Barber, RD Peters, and EC Meslow. 1986. Responses of a black bear population to a changing environment. Int Conf Bear Res and Management 6:57-63.

Lindzey FG and EC Meslow. 1977. Population characteristics of black bears on an island in Washington. J Wildlife Management 41:408-412.

Maessen O, B Freedman, J Svovoda, and MLN Nam. 1983. Resource allocation in high arctic vascular plants of differing growth form. Canadian J Botany 61:1680-1691.

Martinka CJ. 1974. Population characteristics of grizzly bears in Glacier National Park, Montana. J. Mammal. 55:21-29.

McEwan EH. 1968. Growth and development of the barren-ground caribou. II. Postnatal growth rates. Can J Zool 46:1023-1029.

McLellan B. 1989. Dynamics of a grizzly bear population during a period of industrial resource extraction. III. Natality and rate of increase. Can J Zool 67:1865-1868.

McLellan B. 1994. Density-dependent population regulation of brown bears. Pp. 15-24 in M Taylor, Ed. Density-dependent population regulation in black, brown and polar bears. Int Conf Bears Res and Management. Monogr Series No. 3. 43 Pp.

McLellan BN and FW Hovey. 1994. The diet of grizzly bears in the Flathead river drainage of southeastern British Columbia. Can J Zool 73:704-712.

McLellan BN. In press. Maintaining viability of brown bears along the southern fringe of their distribution. Int Conf Bear Res and Management.

Mech LD. 1977. Wolf pack buffer zones as pre reservoirs. Science 198:320-321.

Mech LD. 1986. Wolf numbers and population trend in the central Superior National Forest, 1967-1985. US Dept Agric for Serv Res Pap No. NC-270. 6 Pp.

Mech LD and ME Nelson. 1989. Polygyny in a wild wolf pack. J Mammalogy 70:675-676.

Mech LD. 1994. Biological impacts on wolves of exploitation and control: a response. In First Annual Conference of the Wildlife Society, Albuquerque, NM (abstract).

Mech LD, SH Fritts, and D Wagner. 1994. Minnesota wolf dispersal to Wisconsin and Michigan. US National Biological Survey, Minneapolis, MN. Typewritten manuscript 6 Pp.

Mech LD, TJ Meier, JW Burch, and LG Adams. 1995. Patterns of prey selection by wolves in Denali National Park, Alaska. In LN Carbyn, SH Fritts, and DR Seip, Eds. Ecology and conservation of wolves in a changing world. Canadian Circumpolar Institute, Occasional Papers 35, Edmonton, Alberta.

Meier T, J Burch, LD Mech, and L Adams. 1995. Pack structure and genetic relatedness among wolf packs in a naturally regulated population. Pp. 293-302 in LN Carbyn, SH Fritts, and DR Seip, Eds. Ecology and conservation of wolves in a changing world. Canadian Circumpolar Institute, Occasional Papers 35, Edmonton, Alberta.

Messier F and M Crête. 1985. Moose-wolf dynamics and the natural regulation of moose populations. Oecologia 65:503-512.

Messier F. 1985. Social organization, spatial distribution, and population density of wolves in relation to moose density. Canadian J Zoology 63:1068-1077.

Miller DR. 1980. Wildfire effects of barren-ground caribou wintering on the taiga of northcentral Canada: a reassessment. Pp.84-98 *in* Reimers, E, Gaare E and S. Skjenneberg, Eds. Proceedings of the Second International Reindeer/Caribou Symposium, Direktoratet for vilt og ferskvannfisk, Trondheim, Norway.

Miller F. 1990. Peary caribou status report. Canadian Wildlife Service. West and North Reg Rep. 64 Pp.

Miller SD, EF Becker, and WB Ballard. 1987. Black and brown bear density estimates using modified capture-recapture techniques in Alaska. Int Conf Bear Research and Management 7:23-35.

Miller SD and WB Ballard. 1992. Analysis of an effort to increase moose calf survivorship by increasing hunting of brown bears in southcentral Alaska. Wildlife Society Bulletin 20:445-454.

Miller SD. 1995. Impacts of heavy hunting pressure on the density and demographics of brown bear populations in southcentral Alaska. Alaska Dep Fish and Game. Fed Aid in Wildlife Res Grant W-24-3, Study 4.26. Fairbanks. AK. 28 Pp.

Miller SD, GC White, RA Sellers, HV Reynolds, JW Schoen, K Titus, VG Barnes Jr., RB Smith, RR Nelson, WB Ballard, and CC Schwartz. 1997. Brown and black bear density estimation in Alaska using radio telemetry and replicated mark-resight techniques. Wildlife Monographs 113:1-55.

Miquelle DG, JM Peek, and V Van Ballenberghe. 1992. Sexual segregation in Alaskan moose. Wildlife Monographs 122:1-43.

Molvar EM, RT Bowyer, and V Van Ballenberghe. 1993. Moose herbivory, browse quality, and nutrient cycling in an Alaskan treeline community. Oecologia 94:472-479.

Murie AO. 1944. The wolves of Mount McKinley. Fauna of the national parks of the US. Fauna Ser No. 5. US Gov Print Off, Washington, DC. 238 Pp.

Nagy JA, RH Russel, AM Pearson, MC Kingsley, and CB Larsen. 1983a. A study of grizzly bears on the barren grounds of Tuktoyaktuk Peninsula and Richards Island, Northwest Territories, 1974-1978. Can Wildlife Ser Rep. 136 Pp.

Nagy JA, RH Russel, AM Pearson, MC Kingsley, and BC Goski. 1983b. Ecological studies of the grizzly bear in the Arctic mountains, northern Yukon territory, 1972 to 1975. Can Wildlife Ser Rep. 104 Pp.

Nelson ME and LD Mech. 1986. Relationship between snow depth and gray wolf predation on white-tailed deer. J Wildlife Management 50:90-91.

Oldemeyer JL and WL Regelin. 1987. Forest succession, habitat management, and moose in the Kenai National Wildlife Refuge. Viltrevy (Swedish Wildlife Journal) Supplement 1:163-180.

Osborne TO and others. 1996. Extent, cause, and timing of moose calf mortality in western interior Alaska. Alces 27:24-30.

Oosenbrug S and LN Carbyn. 1982. Winter predation on bison and activity pattern of a pack of wolves in Wood Buffalo National Park. Pp 43-53 *in* FM Harrington and PC Paquet, Eds. Wolves: a worldwide perspective of their behavior, ecology, and conservation. Noyes Publications, Park Ridge, NJ.

Packard JM and LD Mech. 1980. Population regulation in wolves. Pp. 135-150 *in* MN Cohen, RS Malpass, and HG Klein, Eds. Biosocial mechanisms of population regulation. New Haven:Yale Univ Press.

Packard JM, LD Mech, and US Seal. 1983. Social influences on reproduction in wolves. Pp. 78-85 *in* LN Carbyn, Ed. Wolves in Canada and Alaska: their status, biology, and management. Canadian Wildlife Serv. Rep Ser No. 45.

Page RE. 1985. Early calf mortality in northwestern British Columbia. M.Sc. thesis, Univ of Victoria, Victoria, BC 129 Pp.

Parker KL and CT Robbins. 1984. Thermoregulation in mule deer (*Odocoileus hemionus hemionus*) and elk (*Cervus elaphus nelsoni*). Canadian J Zoology 62:1409-1422.

Pastor J. 1988. Moose, microbes, and the boreal forest. Bioscience 38:770-777.

Pearson AM. 1975. The northern interior grizzly bear. Can. Wildlife. Serv. Rep. Ser. No. 34. 86 Pp.

Peek JM. 1974. Initial response of moose to a forest fire in northeastern Minnesota. Am Midland Naturalist 91:435-438.

Pegau RE. 1970. Succession in two exclosures near Unalakleet, Alaska. Canadian Field-Naturalist 84:12-25

Pegau RE. 1975. Analysis of the Nelchina caribou range. Biol Pap Univ Alaska special report 1.

Peterson RO and RE Page. 1983. Wolf-moose fluctuation in Isle Royale National Park, Michigan USA. Annales Zoologici Fennica 74:251-253.

Peterson RO, JD Woolington, and TN Bailey. 1984. Wolves of the Kenai Peninusla. Alaska Wildlife Monographs 88. 52 Pp.

Peterson RO and RE Page. 1988. The rise and fall of Isle Royale wolves, 1975-1986. J Mammalogy 69:89-99.

Peterson RO. 1977. Wolf ecology and prey relationships on Isle Royale. National Park Service Scientific Monograph 11. 210 Pp.

Pianka ER. 1978. Evolutionary Ecology. 2nd Ed. Harper & Row, New York.

Pimlott DH, JA Shannon, and GB Koleosky. 1969. The ecology of the timber wolf in Algonquin Provincial Park. Ontario Dept Lands and For Res Rep (Wildl) No. 87. 92 Pp.

Pritchard GT and CT Robbins. 1990. Digestive efficiencies of grizzly and black bears. Can J Zool 68:1645-1651.

Pruitt WO. 1981. Application of the varrio snow index to over wintering North American barren-ground caribou (*Rangifer tarandus arcticus*). Canadian Field-Naturalist 95:363-365.

Pulliam HR and NM Haddad. 1994. Human population growth and the carrying capacity concept. Bull. of the Ecological Society of America 75:141-157.

Rausch RA. 1969. A summary of wolf studies in southcentral Alaska, 1957-1968. Trans North Am Wildlife and Nat Resour Conf 34:117-131.

Regelin WL, CC Schwartz, and AW Franzmann. 1987. Effects of forest succession on the nutritional dynamics of moose forage. Viltrevy (Swedish Wildlife Journal). Supplement. 1:247-264.

Reimers E. 1980. Activity pattern; the major determinant for growth fattening in Rangifer. Pp. 466-474 in E Reimers, E Gaare, and S Skjenneberg, Eds. Proceedings of the Second International Reindeer/Caribou Symposium, Direktoratet for vilt og ferskvannfisk, Trondheim, Norway.

Reiner DC. 1996. Estimating bear habitat quality in southeastern British Columbia by measuring changes in bear body condition. M.S. thesis, Univ Montana, Missoula, MT. 81 Pp.

Renecker LA and RJ Hudson. 1986. Seasonal energy expenditures and thermoregulatory responses of moose (*Alces alces*). Canadian J Zoology 64:322-327

Reynolds HV and TA Boudreau. 1990. Effects of harvest rates on grizzly bear population dynamics in the northcentral range. Alaska Dep Fish and Game. Div Wildlife Cons. Fed Aid in Wildlife Restoration Research progress report Project W-23-3, study 4.19.

Reynolds HV. 1994. Effects of harvest on grizzly bear population dynamics in the northcentral Alaska Range. Alaska Dep Fish and Game. Fed Aid in Wildlife Res Grant W-24-3, Study 4.26. Fairbanks, AK. 28 Pp.

Robbins CT, S Mole, AE Hagerman, and TA Hanley. 1987. Role of tannins in defending plants against ruminants reduction in dry matter digestion. Ecology 68:1606-1615.

Rogers LL. 1987. Effects of food supply and kinship on social behavior, movements, and population growth of black bears in Northeastern Minnesota. Wildlife Monographs 97:1-72.

Rosenthal GA and DH Janzen, Eds. 1979. Herbivores: their interaction with secondary plant metabolites. Academic Press, New York, NY.

Rothman RJ and LD Mech. 1979. Scent-marking in lone wolves and newly formed pairs. Animal Behavior. 27:750-760.

Schlegel M. 1983. Factors affecting calf survival in the Lochsa elk herd. Job Completion Report, Federal aid project W-1600-R, Study 1, Job 3. Idaho Dept of Fish and Game, Boise, ID.

Schoen JW and LR Beier. 1990. Brown bear habitat preferences and brown bear logging and mining relationships in southeast Alaska. Alaska Department of Fish and Game, Fed Aid in Wildlife Restor Final Report. Project W22:1-6 and W 23:1-3. Juneau, AK 90 Pp.

Schwartz CC, WL Regelin, and AW Franzmann. 1988. Estimates of digestibility of birch, willow and aspen in moose. J Wildlife Management 52:33-37.

Schwartz CC and AW Franzmann. 1991. Interrelationship of black bears to moose and forest succession in the northern coniferous forest. Wildlife Monographs 113:1-58.

Scotter GW. 1967. Effects of fire on barren-ground caribou and their forest habitat in northern Canada. Trans North Amer Wildlife Nat Res Conf 32:246-259.

Scotter GW. 1995. Influence of harassment by wolves, *Canis lupus*, on barren-ground caribou, *Rangifer tarandus groenlandicus*, movements near the Burnside River, Northwest Territories. Canadian Field-Naturalist 109:452-453.

Seal US, LD Mech, and V VanBallenberghe. 1975. Blood analyses of wolf pups and their ecological and metabolic interpretation. J Mammalogy 56:64-75.

Seber GAF. 1982. The estimation of animal abundance and related parameters, 2nd ed. Griffin, London. 654 Pp.

Seip DR. 1992. Factors limiting woodland caribou populations and their interrelationships with wolves and moose in southeastern British Columbia. Canadian J Zoology 70:1494-1503.

Seip DR and K Brown. 1996. Introduction to the population ecology of North American caribou. Rangifer 9:11-12.

Selkregg LL. 1976. 1974-1976 Alaska Regional Profiles: Vol. I Southcentral Alaska, 225 Pp.; Vol. II Arctic, 218 Pp.; Vol. III Southwest, 313 Pp.; Vol. VI Southeast, 233 Pp.; Vol. V Northwest, 265 Pp.; Vol.VI Yukon, 346 Pp. Arctic Information and Data Center, Anchorage, AK.

Sellers RA. 1994. Dynamics of a hunted brown bear population at Black Lake, Alaska. 1993 Annual Progress Report. Alaska Department of Fish and Game. 61 Pp.

Sellers RA, SD Miller, TS Smith, and R Potts. 1993. Population dynamics and habitat partitioning of a naturally regulated brown bear population on the coast of Katmai National Park: 1993 Annual progress report. Alaska Dept of Fish and Game, Div of Wildlife Conservation, Juneau.

Servheen C 1990. The status and conservation of the bears of the world. Eighth Int Conf Bear Res and Management 9:4-16.

Servheen C, FW Hovey, BN McLellan, R Mace, W Wakkinen, W Kasworm, D Carney, T Manley, K Kendall, and R Wielgus. 1994. Report of the northern ecosystems researchers on grizzly bear population trends in the north and south forks of the Flathead, and the Blackfeet Indian Reservation of the NCDE; the Cabinet-Yaak ecosystem; and the Selkirk ecosystem; and future data needs to improve trend estimates. US Fish and Wildlife Service. Missoula, MT.

Sharkey MJ. 1970. The carrying capacity of natural and improved land in different climatic zones. Mammalia 34:564-572.

Skogland T. 1985. The effect of density-dependent resource limitations on the demography of wild reindeer. J Animal Ecology 54:359-374.

Skoog R. 1968. Ecology of the caribou (*Rangifer tarandus granti*) in Alaska. Ph.D. thesis, Univ CA, Berkeley.

Smith TR and MR Pelton. 1990. Home ranges and movements of black bears in a bottomland hardwood forest in Arkansas. Pp. 213-218 *in* LM Darling and WR Archibald, Eds. Bears: their biology and management. International Association for Bear Research and Management, Coeur d'Alene.

Spalding DJ. 1990. The early history of moose (*Alces alces*) : distribution and relative abundance in British Columbia, Victoria, BC. Royal British Columbia Museum.

Stenlund MH. 1955. A field study of the timber wolf, Canis lupus, on the Superior National Forest, MN. MN Dept Conservation Tech Bull 4:1-55.

Stephenson RO. 1975. Wolf Report. Alaska Sept of Fish and Game, Juneau.

Stephenson RO, WB Ballard, CA Smith, and K Richardson. 1995. Wolf biology and management in Alaska 1981-92. Pp. 43-54 *in* LN Carbyn, SH Fritts, and DR Seip, Eds. Ecology and conservation of wolves in a changing world. Canadian Circumpolar Institute, Occasional Papers 35, Edmonton, Alberta.

Storer TI and LP Tevis. 1955. California grizzly. University of California Press, Berkeley.

Stringham SF. 1980. Possible impacts of hunting on the grizzly/brown bear, a threatened species. Int Conf Bear Res and Management 4:337-349.

Stringham SF. 1990. Grizzly bear reproductive rate relative to body size. Int Conf Bear Res and Management 8:433-443.

Telfer ES and JP Kelsall. 1984. Adaptation of some large North American mammals for survival in snow. Ecology 65:1828-1834.

Titus K and LR Beier. 1994. Population and habitat ecology of brown bears on Admiralty and Chichagof Islands, Alaska Dept of Fish and Game, Div of Wildlife Cons, Grant W-24-2 study 4.22, Juneau, AK.

US Department of Agriculture. 1994. Wolf depredation on livestock in Minnesota annual update of statistics. USDA, Animal and Plant Health Inspection Service, Animal Damage Control, Grand Rapids, MN.

Valkenburg P and JL Davis. 1986. Calving distribution of Alaska's Steese-Fortymile herd: a case of infidelity? Rangifer (special issue) 1:315-325.

Valkenburg P, JL Davis, JM Ver Hoef, RD Boertje, ME McNay, RM Egan, DJ Reed, CL Gardner, and RW Tobey. 1996. Population decline in the Delta caribou herd with reference to other Alaskan herds. Rangifer (special issue) 9:53-62.

Van Ballenberghe V and LD Mech. 1975. Weights, growth, and survival of timber wolf pups in Minnesota. J Mammalogy 56:44-63.

Van Ballenberghe V, AW Erickson, and D Byman. 1975. Ecology of the timber wolf in northeastern Minnesota. Wildlife Monographs 43.

Van Ballenberghe V and DG Miquelle. 1993. Mating in moose: timing, behavior, and male access patterns. Canadian J. Zoology 71:1687-1690.

Van Ballenberghe V. 1983. Extraterritorial movements and dispersal of wolves in southcentral Alaska. J Mammalogy 64:168-171.

Viereck LA. 1973. Wildfire in the taiga of Alaska. Quat Res (NY) 3:465-495.

Wayne RK, DA Gilbert, N Lehman, K Hansen, A Eisenhower, D Girman, LD Mech, PJP Gogan, US Seal, and RJ Krumenaker. 1991. Conservation genetics of the endangered Isle Royale gray wolf. Conservation Biology 5:41-51.

Wayne RK, N Lehman, and TK Fuller. 1995. Conservation genetics of the gray wolf. Pp. 399-408 *in* LN Carbyn, SH Fritts, and DR Seip, Eds. Ecology and conservation of wolves in a changing world. Canadian Circumpolar Institute, Occasional Papers 35, Edmonton, Alberta.

White GC. 1983. Evaluation of radio tagging, marking , and sighting estimators of population size using Monte Carlo simulations. Pp. 91-103 *in* JD Lebreton and PM North, Eds. Marked individuals in the study of bird populations. Birkhauser Verlog Basel, Switzerland.

White RG, BR Thompson, T Skogland, SJ Person, DE Russel, D Holleman, and JR Luick. 1975. Ecology of caribou at Prudhoe Bay, Alaska. Biol Pap Univ Alaska, Special Rep 2. Pp.151-201.

White RG and J Trudell. 1980. Habitat preference and forage consumption by reindeer (*Rangifer tarandus tarandus*) and caribou (*Rangifer tarandus granti*) near Atkasook, Alaska. Arctic Alpine Research 12:511-530.

White RG and FA Milan. 1981. Reindeer herds of the Kuskowim, Alaska. Proc Alaska Sci Conf 32:1-10.

Wielgus RB. 1993. Causes and consequences of sexual habitat segregation in grizzly bears. Ph.D. thesis, University of British Columbia.

Wielgus RB and FL Bunnell. 1994. Sexual segregation and female bear avoidance of males. J Wildlife Management 58:405-413.

Wolfe RJ. 1996. Subsistence food harvests in rural Alaska, and food safety issues. Paper presented to the Institute of Medicine. National Academy of Sciences Committee on Environmental Justice, Spokane, WA, August 13.

Wydeven AP, RN Schultz, and RP Thiel. 1995. Monitoring of a recovering gray population in Wisconsin 1979-1991. Pp. 147-156 *in* LN Carbyn, SH Fritts, and DR Seip, Eds. Ecology and conservation of wolves in a changing world. Canadian Circumpolar Institute, Edmonton, Alberta.

Wydeven AP. 1993. Wolves in Wisconsin: recolonization under way. International Wolf 3:18-19.

Zackrisson O. 1977. Influence of forest fires on the North Swedish boreal forest. Oikos 29:22-32.

4

Predator-Prey Interactions

The goal of predator control efforts in Alaska is primarily to increase prey populations for human harvest. Each control effort can be viewed as an experiment whose outcome can be predicted based on ecological theory of predator-prey interactions. It might seem obvious that if predator numbers are reduced, prey numbers would increase and that everything is self-evident without any additional theory. But predator-prey interactions are too complex to assume that fewer wolves automatically means increased prey in a way that wildlife managers feel is beneficial. In particular, if prey are limited by habitat, then predator control may do little good. Also, the cost and political feasibility of predator control depend on how extensively it must be enforced in order to achieve enhanced prey numbers. A policy of predator control would be especially attractive if it allowed for some new equilibrium at which prey and predator populations are both enhanced, because both are valued. Therefore, this chapter begins with an overview of predator-prey theory, which provides the context for the committee's analyses of the results of past wolf and bear control efforts. The chapter concludes with an assessment of the current status of knowledge that can and should be used to determine whether a wolf and/or bear control effort might work in ways that reduce cost and increase political approval.

THEORY OF PREDATOR-PREY INTERACTIONS

Predator-prey theory is one of the oldest and richest branches of theoretical ecology. Its models predict a wide array of results, depending on characteristics of predators, prey, and the environment in which they interact. This section reviews aspects of the theory that are most pertinent to questions about predator control, particularly in Alaska, and discusses their implications for predator management and control in Alaska. Major theoretical results are presented in italics and are followed by a brief discussion of their implications for management.

Oscillations and Stable Levels

Predator and prey populations can either oscillate wildly or persist at relatively stable levels (May 1981).

Theoretical models show that persistence of predators and prey at relatively stable levels (or equilibria) is likely only when prey populations are resource-limited and the prey have a refuge where they are safe from predation. Although the issue of population stability might seem like a topic of theoretical more than immediate practical concern, whether or not a predator-prey system is naturally prone to large fluctuations matters a great deal. In particular, if Alaskan caribou or moose populations naturally fluctuate wildly, the desires of hunters for stable prey populations and reliable harvests year after year might be contrary to natural population dynamics and an inordinate amount of human intervention might be required to achieve stable harvest levels.

Removal of Predators from A Plant-Herbivore-Predator Interaction System

The removal of predators from a plant-herbivore-predator interaction system can either stabilize or destabilize herbivore population dynamics (May 1991).

Predator control is in effect a predator-removal experiment. In addition to the direct results of such a manipulation (fewer wolves or bears), dramatic alterations in the character of prey population dynamics can be triggered. Models indicate that predator removal can destabilize or stabilize herbivore population dynamics or have no effect whatever. The result depends subtly on the rate at which the environmental carrying capacity for the herbivores changes relative to the rate at which predator populations change (Crawley 1983). If vegetation changes rapidly and predators exert their effects during substantial vegetation changes, the presence of predators typically stabilizes prey populations. In other words, long-lived (relative to the speed at which plant resources available to herbivores change) and starvation-tolerant or generalist predators can stabilize numbers of their prey. If wolves in Alaska act in this manner, even though the removal of wolves might yield more prey over the short run, substantial reductions in wolves could increase rather than decrease fluctuations in moose and caribou numbers over a long time span.

Alternative Stable States

Two alternative stable states can exist in predator-prey systems: a lower equilibrium corresponding to very low prey and predator populations, and a higher equilibrium corresponding to high predator and prey populations (with the prey close to their carrying capacity).

This idea received much attention in the 1970s from mathematical biologists, many of whom were inspired by so-called catastrophe theory (Walters and others 1975; Ludwig and others 1978). According to this theory, each equilibrium state "attracts" neighboring population trajectories, so that a shift from one equilibrium to the other can be abrupt, and can occur simply if densities of predators or prey are temporarily altered, without the need for sustained management. The "predator pit" alluded to by wildlife biologists is a verbal version of these models in which it is assumed there is a low density equilibrium for these systems separated from a high density, relatively stable equilibrium. Prey populations are in the "predator pit" when their yearly losses to predation are greater than their yearly population gains (Seip 1995). Prey populations in the predator pit will decline to the lower equilibrium. The key requirement for 2 different equilibria is a functional response on the part of the predators such that the per capita risk of being taken by a predator actually increases with increasing prey density in the neighborhood of the lower equilibrium but decreases in the neighborhood of the higher equilibrium. The existence of two alternative, relatively stable states would provide a strong argument for predator control because the elimination or reduction of predators for a brief time period could kick the system into the higher equilibrium—that is, an equilibrium with more predators and more prey. In this context, equilibrium does not mean constant predator and prey densities, but that densities tend to return to the vicinity of the equilibrium if they are caused to deviate substantially from it.

Using Regression Analysis to Estimate Growth Rates

If a simple regression analysis is used to ask what controls prey populations in a predator-prey system, the factor that explains the greatest proportion of the variance in prey population growth rates depends largely on where "noise" enters the system, and not on what actually controls the dynamics (Boyce and Anderson 1997).

Simulations of predator-prey dynamics with environmental variation entered in different places (for example, in herbivore carrying capacity or in the feeding rate of wolves) have shown that regression analyses of annual increments in wolf and ungulate populations can be very misleading. The pattern of random environmental variation might dictate the outcome of the regressions much more than do the actual linkages between predators and prey. That result is disturbing if one is going to rely on simple regressions and correlations to test hypotheses about "what controls caribou or moose populations." In particular, the fact that calf:cow ratios or that moose and caribou numbers are negatively correlated with numbers of wolves, does not mean that wolves are the primary determinant of and regulating force in moose and caribou populations. In other words, "correlation does not prove causation." Unfortunately, the limitations of correlative evidence are often forgotten during discussions of population dynamics. Those limitations are especially severe when one examines highly complex interactions involving vegetation, herbivores, and carnivores. This does not mean that regression models are uninformative—rather, it means that the best way of fitting data is to fit data that describe dynamics through time, not data that are slices of static patterns (see appendix C for a specific example developed by the committee).

INTEGRATING THEORY AND DATA

Models of predator-prey interactions make it clear that possible outcomes of predator control experiments are highly varied and are likely to depend on the particular conditions under which control was carried out. Therefore, long-term management could be improved if a more solid understanding of wolf-caribou or wolf-moose interactions were available. For instance, a definitive demonstration of the existence or absence of alternative stable states would be a key piece of information. More generally, the type of vegetation and how it changes through time determine the impact of predators on herbivorous prey. Calculating simple correlations between hypothetical driving variables constitutes a weak form of analysis. In a dynamic system such as this, more is gained by fitting population dynamic models to the data as opposed to simply seeing what is correlated with what (see appendix C).

The practical implications of predator-prey theory are that:

1. Correlative studies have limited abilities to determine causal relationships.
2. The interactions between prey and their plant resources need to be understood.
3. The task of identifying which "model" describes a particular situation is technically challenging.

Because of the above challenges, analysis of predator control in Alaska is a major scientific task. Moreover, it is a task that can gain only limited guidance from scientific studies elsewhere, inasmuch as these same technical challenges have thwarted resource management throughout the world. In short, a real understanding of predator management and its consequences in Alaska will require state-of-the-art science carried out in Alaska. In the end, however, perfect understanding is unattainable and we must learn to live with and base our actions on incomplete knowledge and imperfect predictive abilities.

From a management perspective, three important questions are (1) can predators maintain ungulate populations at densities well below carrying capacity, (2) if so, how long can predators keep ungulate population densities low, (3) if prey populations increase to the environmental carrying capacity, how long will they remain there after predator control ceases? To determine whether predators can depress ungulate populations, factors affecting recruitment (birth and immigration) and loss (death and emigration) must be evaluated. Each of these, in turn, depends on environmental factors such as characteristics and distribution of forage plants, weather, parasitism, disease, the availability of mates, and predation by wolves, bears, and humans. Those factors can interact with the density of the population in such a way that birth rates might decrease but predation rates may increase as population density increases.

The number of prey killed by predators depends on the number of predators and the number of prey killed by each predator; both of these factors are related to the abundance of prey. The change in the density of predators in relation to prey density is called the *numerical response*. The change in the number of prey killed by an individual predator in relation to prey density is called the *functional response*. Although functional and numerical responses of predators are often linear over restricted ranges of prey density, they are strongly nonlinear when one considers a wide range of prey densities.

Refugia from predators or other anti-predator behavior often reduce both the numerical and functional response of predators at very low prey densities. Both responses, however, usually increase rapidly with moderate but increasing prey densities, but the functional response levels off at higher prey densities due to satiation while the numerical response might level off at high prey densities due to the social structure of the predators. The product of the numerical and functional response equals the predation rate, and it generally increases with increasing prey density. When this happens, prey populations are likely to grow rapidly at very low population densities but their growth rates should slow down at higher population densities. However, under some conditions, the responses of predators are greatest at moderate prey densities and decline thereafter. If such a pattern exists, predators can prevent prey populations from increasing when they are at low densities, but if prey are able to increase to higher densities, their densities can increase beyond the point where predators exert effective control. The result is a system with relatively stable low and high prey densities, separated by intermediate densities at which the predation rate is too low to prevent the prey from increasing to a higher equilibrium; that is, where predation is relatively unimportant (figure 4.1). The rationale often given to justify wolf control is that if humans reduce the numbers of wolves, a low-density moose or caribou population might be released from its low equilibrium and allowed to stabilize at a higher-density equilibrium (Gasaway and others 1992). Haber (1977) has suggested the existence of a low equilibrium predator pit at 0.02-0.20 moose/km^2

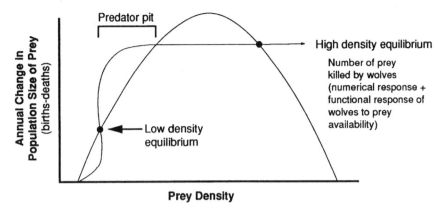

FIGURE 4.1 Multiple equilibria. Multiple equilibria can exist for predators and their prey predator-caused mortality increases with increasing prey density up to some point, and then decreases.

and an upper equilibrium at 0.4 moose/km^2. However, there is little evidence of stable high-density equilibria in nature (Messier 1994; Seip 1995).

Van Ballenberghe (1987) reviewed the literature and concluded that 2-predator / single prey systems are more likely to be stable at low densities than are 1-predator / one prey systems. Thus, although wolf predation alone could limit the size or growth of a prey population (Ballard and others 1987; Larsen and others 1989; Peterson and others 1984a,b), the presence of a second predator, such as brown bears, can favor a low-density equilibrium (Messier and Crête 1985). Mortality during summer often removes 50% of annual calf production in ungulates; much of this predation is by bears, but its effects on population dynamics are not clear, because increased summer survival might be compensated for by decreased winter survival. Most researchers do not follow calves through the winter, even though population growth rate is more sensitive to mortality in older animals (Linnell and others 1995).

Thus, in combination, wolves and bears can drive prey populations to low population densities. However, from a management perspective, the more important question is whether wolf predation can *maintain* prey populations at low densities for long periods (Sinclair 1989). Existing evidence on the point is inconclusive. For example, Messier (1994) reviewed the literature and concluded that wolf predation can be density-dependent at low moose densities (not more than 0.65 moose/km^2) but is often inversely density dependent at higher moose densities. If this were generally true, predators could depress prey populations for extended periods and control should allow a prey population to increase to, and remain for some time at, a higher density determined by food availability (Walters and others 1981), rather than by predation (Messier 1994;

Sinclair 1989). But Boutin (1992) reviewed the same literature and concluded that the evidence that predation is a major limiting factor in most moose populations is not convincing.

Differences in interpretation are possible because the supporting data are sparse. Not enough is known about the patterns of numerical and functional responses of wolves and the capacity of moose to sustain losses at different densities and environmental conditions. The cost of predator control programs would be reduced and their political acceptability would be increased if prey populations remained at high densities for a long time after predator control ceased and predator population densities rebounded to their pre-control levels. Factors that could result in persistent high prey densities include territorial behavior and intraspecific strife, which might set an upper limit on wolf numbers (Messier 1994). Alternatively, wolves that normally prey on moose might switch to caribou when they are particularly abundant, but fail to return to their earlier levels of predation on moose when caribou decrease in numbers (Dale and others 1994; Gasaway and others 1992). In his study on the numerical and functional responses of wolves in the Yukon, Hayes (1995) found that substantial changes in caribou distribution, or snowshoe hare abundance had little effect on wolf predation on moose. Before wolf-bear-prey interactions can be fully understood, more studies on functional and numerical responses of wolf populations to various combinations of prey (caribou, sheep, beavers, and snowshoe hares), of the behavior of bear populations, of the nutritional condition of the adult prey killed by wolves, and of the conditions under which wolves cause substantial calf mortality in prey will be needed (Messier 1994). However, it is already clear that no single pattern dominates those interactions. Variations in weather, habitat conditions, and behavior of predators and prey guarantee that outcomes will be varied, difficult to predict, and difficult to interpret.

REFERENCES

Ballard WB, JS Whitman, and CL Gardner. 1987. Ecology of an exploited wolf population in south-central Alaska. Wildl Monogr 98:1-54.

Boutin S. 1992. Predation and moose population dynamics: a critique. J Wildl Manage 56:116-127.

Boyce MS and RM Anderson. 1997. Evaluating the role of carnivores in the greater Yellowstone ecosystem. *In* Carnivores in ecosystems. Yale University Press, New Haven.

Crawley MJ. 1983. Herbivory, the dynamics of animal-plant interactions. Studies in Ecology, v. 10, University of California Press, Berkeley, CA.

Dale BW, LG Adams, and RT Bowyer. 1994. Functional response of wolves preying on barren-ground caribou in a multiple prey ecosystem. J Anim Ecol 63:64-652.

Gasaway WC, RD Boertje, DV Grangaard, DG Kellyhouse, RO Stephenson, and DG Larsen. 1992. The role of predation in limiting moose at low densities in Alaska and Yukon and implications for conservation. Wildl Monogr 120:1-59.

Haber GC. 1977. Socio-ecological dynamics of wolves and prey in a subarctic system. Ph.D. thesis, University of British Columbia, Vancouver.

Hayes RD. 1995. Numerical and functional response of wolves, and regulation of moose in the Yukon. M.Sc. thesis, Simon Fraser University, Burnaby, British Columbia.

Larson DG, DA Gauthier, RL Markel, and RD Hayes. 1989. Limiting factors on moose population growth in the southwest Yukon. Yukon Fish and Wildl Br Rep. Whitehouse. 105 Pp.

Linnell JDC, R Aanes, and R Andersen. 1995. Who killed Bambi? The role of predation in the neonatal mortality of temporal ungulates. Wildl Biol 1:209-223.

Ludwig D, DD Jones, and CS Holling. 1978. Qualitative analysis of insect outbreak systems the spruce budworm and forest. J Anim Ecol 47: 315-332.

Messier F and C Crête. 1985. Moose-wolf dynamics and the natural regulation of moose populations. Oecologia 65:503-512.

Messier F. 1994. Ungulate population models with predation: a case study with the North American moose. Ecol 75:478-488.

Peterson RO, JD Woolington, and TN Bailey. 1984a. Wolves on the Kenai Peninsula, Alaska. Wildl Monogr 88:1-52.

Peterson RO, RE Page, and KM Dodge. 1984b. Wolves, moose, and the allometry of population cycles. Science. 224:1350-1352.

Seip DR. 1995. Introduction to wolf-prey interactions. *In* LN Carbyn, SH Fritts, and DR Seip, Eds. Ecology and conservation of wolves in a changing world. Canadian Circumpolar Inst, Univ Alberta, Edmonton, 1995.

Sinclair ARE. 1989. Population regulation in animals. *In* JM Cherrett, Ed. Ecological concepts: the contribution of ecology to an understanding of the natural world. Blackwell Scientific Publications, Oxford.

Van Ballenberghe V. 1987. Effects of predation on moose numbers: a review of recent North American studies. Viltrevy (Swedish Wildlife Research) Supplement 1:431-460.

Walters CJ, R Hilborn, and R Peterman. 1975. Computer simulation of barren-ground caribou dynamics. Ecol Modelling. 1:303-315.

Walters CJ, M Stocker, and GC Haber. 1981. Simulation and optimization models for a wolf-ungulate system. Pp. 317-337 *in* CW Fowler and TD Smith, Eds. Dynamics of large mammal populations. John Wiley and Sons, NY.

5

Wolf and Bear Management: Experiments and Evaluations

INTRODUCTION

In addition to many studies on the population ecology of wolves, bears, moose, and caribou, a number of experiments have been conducted in Alaska and elsewhere, in which wolf and/or bear numbers were reduced or their behavior changed, and responses of caribou and/or moose populations were monitored. These control activities have been targeted to specific areas that cover a relatively small part of the state. In most of these experiments, wolves were killed, but some used translocation or diversionary feeding. These experiments provide the best data with which to evaluate the biological basis of control as a management tool to increase ungulate numbers (Theberge and Gauthier 1985; Boutin 1992). In this section, the committee analyzes and evaluates these control experiments.

Although, the primary goal of wolf control and/or bear management in Alaska is to increase the availability of moose and caribou for human harvest, most management actions have not been monitored to directly assess whether, in fact, this goal was achieved. Instead wildlife managers have relied on less-expensive and short-term measurements, such as changes in birth rates (calf:cow ratios) or changes in adult population sizes, to assess the results of predator management experiments. The following cases are presented in order from the most to the least direct measurements of whether predator management resulted in increased human harvest of moose or caribou (table 5.1; figure 5.1).

TABLE 5.1   Predator Reductions Discussed in Chapter 5

Method of Predator Reduction*	Duration (Years)		Prey Response Measured
	Wolves	Bears	
Air-assisted			
East-central AK (GMU 20A)[a]	7	Not done	Calf survival, adult mortality
Finlayson, Yukon[b]	6	Not done	Calf:cow ratios, adult mortality, population densities, hunting success
Southwest, Yukon[c]	5	5	Population densities, survival rates
Aishihik, Yukon[d]	4	Not done	Calf:cow ratios
Northern BC[e]	10†	Not done	Calf:cow ratios; population densities
Québec[f]	4	3	Calf:cow ratios
East-central AK (GMU 20E) [g]	3	Not done	Calf:cow ratios; calf mortality
South-central AK (GMU 13)[h]	3‡	1	Calf:cow ratios
Ground-based			
Kenai Peninsula, AK[i]	3	Not done	Population densities
Vancouver Island, BC[j]	4	Not done	Hunting success
East-central Saskatchewan[k]	Not done	1,1 (different areas)	Calf:cow ratios

*See text for explanation of grouping by methods. Several experiments involved multiple periods of predator reduction. Primary sources: a, Boertje and others 1995; b, Farnell and Hayes 1992; Farnell and others, in preparation; Larsen and Ward 1995; c, Larson and others 1989a, 1989b, Hayes and others 1991; d, Hayes 1992, Yukon Fish and Wildlife Branch 1994, 1996; e, Bergerud 1990, Elliott 1986a, 1986b, 1989; f, Crête and Jolicoeur 1987; g, Gasaway and others 1992; h, Ballard 1991; i, Peterson and others 1984; j, Archibald and others 1991; k, Stewart and others 1985.

†The areas in which wolf populations were reduced spanned a 10-year period, but was not done in the same place every year.

‡A combined aerial shooting and poisoning program is also described under this case study.

AIR-ASSISTED WOLF CONTROL

East-central Alaska (Delta, GMU 20A)

The best documented and most successful example of wolf control in Alaska was conducted from 1976 to 1982 in Game Management Unit 20A (GMU 20A), south of Fairbanks. The 17,000 km² study area included 7300 km² in the poorly drained lowlands of the Tanana Flats and 9700 km² to the south in the foothills

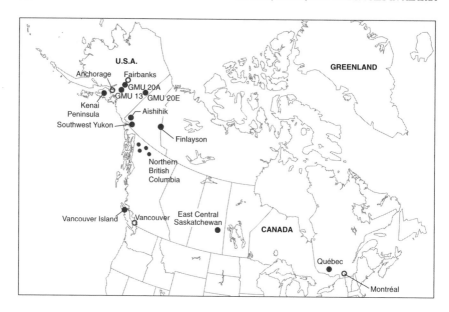

FIGURE 5.1   Locations of case studies discussed in this chapter are indicated by gray dots. Predator control in the northern British Columbia case study was carried out over several distinct areas, as indicated by the 4 smaller dots.

and mountains of the Alaska Range (Gasaway and others 1983; Boertje and others 1995; figure 5.2).

Before wolf control was begun in 1976, the average age of the moose population was very low. Brown and black bear populations were judged to be low. Between 1965 and 1975, moose and the Delta caribou herd in this area were overharvested by humans (Gasaway and others 1983). A record snowfall in the winter of 1970-71 had caused substantial mortality of moose. Local trappers were taking approximately 20% of the wolves annually, but this harvest was not enough to reduce the population of wolves in the area. The caribou hunting season was closed in beginning in 1973.

From 1976 until 1982, a 7-year air-assisted wolf control program was conducted by both ADFG staff and private hunters. Each year during this period, the wolf population was reduced to 55-80% below pre-control numbers. Regular harvest of wolves by private trappers continued throughout the period of wolf control. In 1976 there was an estimated 14 wolves per 1000 km^2; in 1982 at the end of the control period there was an estimated 8 wolves per 1000 km^2.

During the 7 years of wolf control, survival of moose calves and yearlings increased and mortality of adults, especially middle-aged and old adults, declined. The moose population increased from 183 to 481 per 1000 km^2, a mean

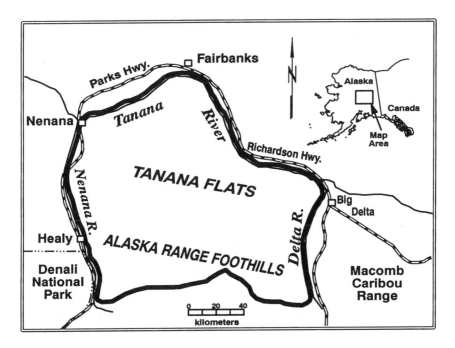

FIGURE 5.2 Study area in interior Alaska (GMU 20A) where wolves were controlled during 7 winters, 1975-76 through 1981-82. Wolves were controlled in a 10,000 km^2 portion of the 17,000 km^2 study area during winters 1993-94 and 1994-95.

annual rate of increase of 15% . After the wolf control program was terminated, the moose population continued to increase for 12 more years, reaching 1020 moose per 1000 km^2 by 1994, a mean annual increase of 5%.

During the period of wolf control (1976-1982), calf survival in the Delta caribou herd increased and adult mortality declined, contributing to an average annual rate of 16% which then continued to increase at 6% for 7 years after wolf control ended (1983-1989). The peak density was 890 caribou per 1000 km^2. The subsequent decline in caribou in the early 1990s coincided with several severe winters that ended the previous 20 years of mild winters. Two adjacent, low-density caribou herds also declined during this period.

By 1985, 3 years after the control programs were ended, wolf numbers had recovered to near precontrol levels in most of the area. After ADFG terminated this wolf control program, private hunters continued to take up to 25% of the autumn wolf population each year. This level of harvest presumably had some impact on the wolf population, even though census data did not reveal a decline in the number of wolves (Peterson and others 1984; Gasaway and others 1992). Human harvest of moose during the 20-year period was restricted to maintain at least 30 males for every 100 females. Board of Game regulations prohibited the harvesting of female moose. Caribou harvest was kept to less than 6% during the 20-year period, except for 1983-1986 when it was 11%. With the decline of caribou populations in the early 1990s, legal human harvest of caribou was reduced and then eliminated.

Documentation of the responses of wolves, moose, and caribou in GMU 20A was based on intensive sampling by ADFG personnel and analysis of information from hunter moose harvest reports. Data for wolves were based on aerial surveys; information from local trappers, hunters and pilots, and from tracking radio-collared wolves (Gasaway and others 1992). Estimates of moose abundance were based on data from aerial surveys (Gasaway and others 1992) and estimates of harvest rates. Birth rates of radio-collared female moose and caribou were compared to estimates of birth rates in other surveys. Ten-month-old calves were weighed after they had received immobilizing doses of anesthetics. Estimates of caribou abundance were based on aerial photographs, total aerial searches, and radio-search techniques. Juveniles of known ages were weighed.

Data were also collected from areas where wolf control was not conducted, but they are difficult to interpret. For instance, calf:cow ratios were high for migratory moose that calved in the Tanana Flats (with wolf control), and wintered in the Chena and Salcha areas (no wolf control). But in the Denali, Fortymile, or Tok areas, where wolves were not controlled, calf:cow ratios of moose did not increase in the late 1970s. During this same period, caribou herds in Macomb and Denali were stable.

During and after the 7-year air-assisted wolf control program from 1976 to 1982, moose and caribou populations increased. Further, during the 7 years of wolf control human harvests of moose and caribou were curtailed, and it is difficult to separate the combined effects of wolf control and reduced human harvests. Nevertheless, the positive responses of both moose and caribou lasted longer than in any other control program and no such responses were seen in the untreated areas.

In the winters of 1989-90, 1990-91, and 1992-93, snow reached critical depths and yearling moose survival was low. The growing season in 1992 was particularly short, and no twin moose were reported the following spring. Except for 1992, caribou birth and recruitment were reduced in GMU 20A and in the comparison Denali and Macomb herds. The Delta caribou herd declined from 1989-93 at an annual average rate of 0.78. In contrast, the moose population apparently increased between early winter 1988 and 1994.

During the winters of 1993-94 and 1994-95, a ground-based wolf control program was conducted in the Tanana Flats to determine whether wolf control would reverse the decline of the Delta caribou herd and allow their numbers to increase again. Before control there were 262 wolves in the area. This was reduced by 62% in 1993-94 and by 56% in 1994-1995. The Delta caribou herd appeared to increase in numbers after the winter of 1993-94. However, 2 years of ground-based wolf control in a 600 km^2 area in the early 1990s apparently had no effect on caribou calf:cow ratios.

Finlayson, Yukon Territory

The 14-year Finlayson wolf reduction experiment included 6 years of wolf reductions and 8 years when populations were monitored in the absence of reductions. It is a very important experiment because although it was not conducted with an appropriate experimental design, it provides the best available data to test the two equilibrium or "predator pit" hypothesis. In addition, wolves were reduced to lower levels for more years than most other experiments and the population ecology and behavior of wolves was studied in detail after control ceased.

The Finlayson Caribou Herd (FCH) lives on a 23,000 km^2 area in the south-central Yukon Territory (figure 5.3). FCH wintered in a complex of hills and small plateaus separated by a several broad, glaciated valleys. The area is typical northern boreal forest with black and white spruce, lodgepole pine, and occasional stands of aspen. The large mammal system includes caribou, moose, 200-300 mountain goats, less than 100 Dall's sheep, a few mule deer, wolves, brown bears, black bears, coyotes, wolverine, lynx, and red fox. Beaver and Arctic ground squirrels were common and the cyclic snowshoe hare populations declined in 1983 and again in 1992. An all-season highway that bisects the FCH winter range facilitates hunting, particularly during the winter when the caribou distribution straddles the road. Another road provides summer and fall access for hunters to the northern portion of the area.

Conditions Before Wolf Reduction

In 1982 the caribou population was crudely estimated to be between 2,000 and 2,500 animals. In October 1982, there were 17 calves/100 cows, a ratio thought to be insufficient to support the estimated harvest of 250 adults per year. In October 1983, a ratio of 34 calves/100 cows was recorded, giving a pre-treatment average of 25.5 calves/100 cows for those two years. In 1982 and 1983, the annual mortality rate of radio-collared adult caribou was 28% (n=5/18). In the winter of 1983, an estimated 215 wolves lived in the study area (9.3/1000 km^2) with an average pack size of 9.6 wolves.

Wolf Reduction

Wolves were systematically reduced by aerial shooting in late winter, with an effort to remove entire packs. In the winter of 1983, 49% of the 215 wolves that lived in the study area were shot. Over the next 5 winters, from 80 to 85% of the wolves were shot each year. Of the 454 wolves removed, 77% were shot from

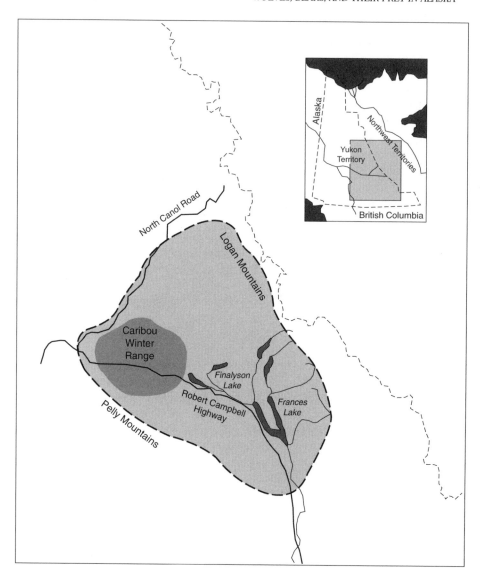

FIGURE 5.3   Location of the Finlayson study area in the Yukon, Canada.

a helicopter. In addition, hunter harvest of caribou was reduced from about 250 per year prior to 1983 to about 25 thereafter.

Response When Wolf Numbers Kept Low

In 1986 there were estimated to be 3073 ± 333 (90% confidence limits) caribou. In 1990, the year following the end of 6 years of wolf control, there were 5950 ± 1055 (90% confidence limits) caribou, an average rate of increase of 18% per year. Mortality rate of radio-collared adult caribou, which was 24% in 1983 after a 49% wolf reduction, decreased further to an average of 11% between 1984 and 1987. The pre-treatment average of 25.5 calves/100 cows almost doubled to 50.2 (42, 50, 45, 55, 47, 62) calves/100 cows during the years of wolf reduction. Two neighboring untreated caribou herds had between 15 and 28 calves/100 cows in 1985 and 1986 when the FCH had 45 and 50 calves/100 cows.

Moose were not the focal prey species of this project and few pre-treatment data are available. Censuses in 2 portions of the treatment area, North Canol (NC) and Frances Lake (FL), were conducted in 1987 during the fourth year of the wolf control program. At that time the moose population was estimated to be 516 in NC and 741 in FL, and there were 67 and 66 calves/100 cows respectively. In 1991, the number of moose had increased to 950 in NC and 1475 in FL for rates of increase of 16% and 18% per year. The number of days it took a hunter to kill a moose decreased from 26 in 1979-1984 to 18 in 1985-1991 in NC. In FL, the number of hunter days required to kill a moose decreased from 31 to 23.

Response When Wolf Control Stopped

Wolves increased from 29 survivors at the end of wolf control (March 1989) to 240 in March 1994, and to 260 in 1996, 17% higher than the pre-control level of 215 wolves. The number of packs increased from 7 in 1989 to between 26 and 28 after 1991. Mean number of wolves per pack increased from 4.4 in 1990 to 7.8 in 1994 and 9.3 in 1996; almost identical to the pre-treatment pack size average of 9.6 wolves per pack.

Detailed monitoring of wolf predation rates on moose demonstrated that wolves in small packs had greater kill rates per wolf than wolves in large packs. In addition, wolves were efficient predators when moose were at low densities. Kill rate was density-independent when there were between 250 and 430 moose/1000 km^2 for all sizes of wolf packs. This result is important because it demonstrates that in Finlayson study area, when moose densities are low (less than 250 moose/1000 km^2) wolves can maintain the population at low densities (estimated to be at 120 moose/1000 km^2). Whether or not there is also a high equilibrium remains unknown, but the efficiency of wolf predation suggests it is unlikely for the interior of the Yukon.

After 1989 and the end of wolf control, caribou calves/100 cows gradually declined from an average of 50.2 when wolves were low to about 44, 32, 20, and 30 in the following 4 years; about the same as in adjacent herds where there was no wolf control. The caribou population decreased from about 5950 ± 1055 to 4550 ± 550 for an average rate of decrease of 5% per year.

A moose census of both NC and FL census areas in 1996 found that the number of calves/100 cows had decreased to 29 and 30 from the 1991 ratios of 52 and 44 respectively. The estimated number of moose in NC had decreased from 1001 in 1991 to 810 in 1996, an average annual rate of decrease of 4%. Similarly, the estimated number of moose in FL decreased from 1475 in 1991 to 1320, an average rate of decrease of 2%.

To summarize, although hunting was reduced during wolf control in Finlayson, the responses of both moose and caribou were clear. The removal of approximately 85% of the wolf population for 6 years and reducing the harvest rate appeared to result in an increase in adult caribou survival, an increase in calves/100 cows of both caribou and moose, and increases of both caribou and moose populations at a rate of approximately 18% per year. Once wolf control ceased, the wolves rapidly increased and pack size recovered in 4 years. Wolf predation rate on moose was independent of moose density, suggesting that a low moose/wolf equilibrium existed. Seven years after wolf control stopped and 3 years after wolves recovered, moose and caribou numbers appeared to be decreasing. This decrease suggests either that a higher equilibrium does not exist in the area, or, if it does exist, it is at a higher density than the ungulates reached during the wolf control years. However, because there was a great increase in mineral exploration in the area in the 1990s, human activity may have prevented moose and caribou from attaining even higher population densities.

Southwest Yukon

In the early 1980s hunter demand for moose in southwestern Yukon exceeded the supply; moose populations either declined or remained stable at low numbers. To assess the effects of brown bear and wolf predation on limiting moose populations, the causes and rates of moose mortality were documented between 1983 and 1987. The effects of hunting, weather, moose reproduction, forage availability, and emigration on moose numbers were also evaluated. The study included 3 experimental wolf reduction areas: Haines Junction (4890 km^2), Rose Lake (6310 km^2) and Lorne Mountain (1020 km^2), and 2 areas in which wolves were not manipulated: Teslin (2580 km^2) and the Auriol Range (1190 km^2) (figure 5.4). Survival rates and causes of mortality of adult moose and calves were compared among experimental and control areas.

Conditions Before Wolf Reduction

Before wolf reduction, moose populations in two of the three experimental

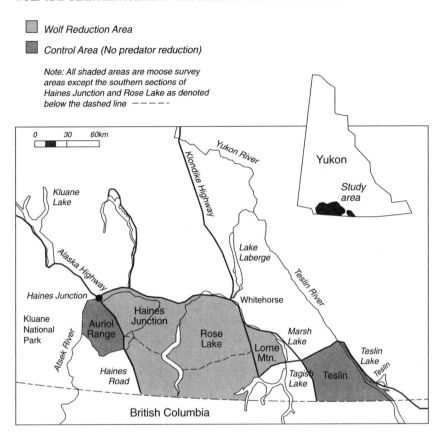

FIGURE 5.4   Location of the wolf control study area in the southwest Yukon, Canada.

areas (Haines Junction and Lorne Mountain) appeared to be declining, but were stable or slightly increasing in the third experimental area (Rose Lake). The moose population in one unmanipulated area appeared to be gradually decreasing while in the other unmanipulated area the moose population was stable or slightly increasing. Calf survival varied greatly between the two years before wolf reduction: In autumn 1980 and 1981, respectively, there were 40 and 11 calves/100 cows in Haines Junction, 20 and 26 in Rose Lake, and 37 and 9 in Lorne Mountain. The overall average in the two years before wolf reduction was 23.8 calves/100 cows. In the unmanipulated areas there was an average of 22 calves/100 cows before treatment. Wolf densities in the 3 experimental areas averaged 12.5 wolves per 1000 km^2 before removal and averaged 15 wolves per 1000 km^2 in the areas where wolves were not removed.

Wolf and Bear Reduction

Wolf populations in the experimental areas were reduced over a 5-year period through aerial hunting between the winters of 1982-83 and 1986-87 but the level of reduction varied among years and treatment areas. Reduction of the wolf population by more than 40% was thought to be sufficient to influence moose populations, but this level of reduction was attained for only 3 years in Haines Junction and Lorne Mountain and for 5 years in Rose Lake. Population reductions of 80% were attained for 2 years at Rose Lake and for one year at Lorne Mountain. Liberalized brown bear hunting regulations were used to reduce the number of brown bears in the same experimental areas. The average number of bears removed each year increased from 3 to 6 in Haines Junction and from 2 to 8 in Rose Lake. No bears were removed from Lorne Mountain. The bear population was estimated to be reduced by 7-9% but bears are difficult to census, and this estimate may not be correct.

Response to Wolf and Bear Reduction

Because the size of the moose population was not monitored in either Haines Junction or Lorne Mountain after predator removal, the effects of wolf and bear reduction on moose numbers could be assessed only in the Rose Lake experimental area. The moose population in Rose Lake did not increase significantly after 5 years of reducing the wolf population by more than 66% and 4 years of liberalized bear hunting regulations (607 ± 109 in 1981 and 582 ± 163 in 1982 compared to 717 ± 143 in 1986).

In the 3 treatment areas, the average number of calves/100 cows was 23.8 before predator reduction and 22 during experimental reductions. There was no difference in mean annual female moose survival rate in areas and years with wolf reduction (92%) and areas and times without wolf reduction (88%). However, 8 of the 16 female moose that died of known causes were killed by wolves, 4 by brown bears, 2 by either bears or wolves, and 2 from an unknown predator. Eighty percent of 132 collared moose calves died in their first year, and brown bears killed 58% and wolves 27% of these. The percent of calves that survived to 6 months of age was significantly greater in areas with wolf reductions (31%) than in areas without (21%). Multiple regression analysis also indicated that calf survival was greater when wolf population sizes were low. When wolf reduction years were omitted from this analysis, mean calf survival was significantly higher when maximum snow depth was less than 80 cm (22%) compared to when it was deeper (11%).

Thus, although moose populations increased slightly, reducing wolf numbers by more than 60% for 5 years and liberalizing brown bear harvest in southwest Yukon failed to produce a substantial increase in moose numbers. Wolves were responsible for half of the adult female moose mortalities and 27% of the calf mortalities, but the survival of radio-collared adult moose was similar in untreated and treatment areas. Brown bears caused more than half of the moose calf mortalities. It was concluded that either a greater level of wolf reduction for a longer period of time, or control of both brown bears and wolves would be required to result in substantially increased moose populations. The financial ($1,375,000 CAN) and public relations cost of this program far outweighed the benefits in terms of increased numbers of moose.

Aishihik, Yukon Territory

A well-planned wolf removal and sterilization experiment is underway in the Aishihik area of southwestern Yukon. The Aishihik study area is 20,000 km^2 of rolling hills, located on the eastern boundary of Kluane National Park, in the rain shadow of the massive St. Elias Mountains (figure 5.5). Moose, woodland caribou, and Dall's sheep are important ungulate prey in the area for a variety of predators including wolves, brown bears, black bears, coyotes, wolverine, lynx, and golden eagles. The experimental area is part of the traditional area of 3 First Nations (aboriginal Canadians). Local and traditional knowledge suggested that moose and caribou numbers were lower than they formerly were. Residents of the area believed that the initial cause of the ungulate decline was overhunting, particularly of cows and calves, even though bears and wolves were more abundant than in previous years.

Conditions Before Wolf Reduction

The Aishihik herd declined from about 1,200-1,500 caribou in 1981 (Larsen and others 1989a) to 785 in 1991 and further to 583 in 1992. In October 1990, 1991, and 1992 there was an average of only 15 calves/100 caribou cows (28.8, 8.6, and 7.3 survived to October each year). The adjacent Kluane caribou herd had an average of 20.4 calves/100 cows (31.7, 29.6, and 0.0) over the same 3 years and 6 other untreated herds averaged 24 calves/100 cows. Biologists predicted that the ratio of calves/100 cows would increase to greater than 30-35 if wolves were reduced.

Surveys in two portions of the experimental area indicated that moose populations were decreasing. The pre-treatment densities of 60-114 moose/1000 km^2 were among the lowest moose densities in the Yukon (average is 218 moose/ 1000 km^2). The number of moose calves/100 cows in the area where wolves were controlled was 23 in 1981, 52 in 1990, and 12 in 1992 in one census area and 22 calves/100 cows in 1992 in the second area. After 1993 moose recruitment was reported as percent calves of the entire population including bulls instead of calves/100 cows so results are not comparable. Calves represented 10% of the moose population half a year before the onset of wolf control but averaged 9.7 percent (10, 12, and 7) the same year in other areas. Age class

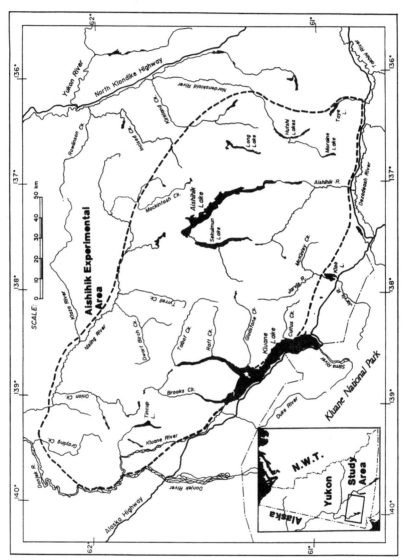

FIGURE 5.5   Location of the Aishihik study area in the Yukon, Canada.

estimates of Dall's sheep in the Aishihik study in 1985-92 averaged 19 lambs/100 ewes; 2 untreated areas averaged 20 lambs/100 ewes during the same period. The wolf population size was estimated at 178 individuals in 1992 (10-11.6 wolves/ 1000 km²) but only 121 in 1993 just prior to the first control session.

Wolf Reduction

In response to the perceived caribou decline and low moose numbers in an area of traditional importance to the 3 First Nations, hunting of caribou was curtailed in 1990 and moose hunting stopped in 1993. From 1993 to 1996 (4 years), 22-64 wolves were removed each winter, leaving 19-22% of the 1992 population size. Fertility control was begun in 1994. Females were sterilized by tubal ligation and males by vasectomy. As of 1997, 5 pairs, one lone male, and one lone female had been sterilized. The ratio of caribou calves to 100 cows increased from an average of 15 when wolf densities were high to an average of 41.5 (39, 38, 42, 47) during wolf control. During the same period the average number of calves/100 cows in the adjacent but untreated Kluane herd increased slightly from 20.4 to 26.4 (19, 22, 33, 32) and the average of 6 other untreated herds increased from 24 to 27 calves/100 cows. The proportion of the moose population that were calves increased from the 10% measured before wolf control to 19% (18, 18, 16, 24). The percentage of calves in the 3 untreated caribou herds increased from 9.7 to an average of 13.5, but data were highly variable. The number of Dall's sheep lambs/100 ewes increased from 19 before wolf control to 27 during control, with the untreated herds increasing from 20 to 26.

Except for the fact that hunting was stopped during the years of wolf control, the Aishihik experiment was well designed. The removal of approximately 76% of the wolf population by aerial wolf control for 4 years and the elimination of hunting resulted in an even greater increase in the number of caribou calves/100 cows than predicted. The response of moose was not as clear, but there was little change in moose calf survival because of high levels of bear predation in the region (Hayes 1992). The proportion of calves in the treatment area grew less than 4% more than in the control areas and there was great variation both within and among control areas. Because mountain sheep are relatively secure from wolf predation in the rugged terrain that they inhabit, Dall's sheep were not expected to be affected by wolf removal. As predicted, the number of Dall's sheep lambs/100 calves showed no noticeable response to wolf removal. Because the experiment ended less than a year ago, changes in ungulate population densities cannot yet be assessed.

Northern British Columbia

Moose, elk, stone sheep and in particular caribou were thought to be declining in northeastern British Columbia in the 1970s and 1980s. In an attempt to

TABLE 5.2   Summary of Wolf Reduction in Northern British Columbia

Year	Location	Area (km2)	Number of Wolves Killed
1978	Horseranch	3,000	22
1979	Horseranch	3,000	25
1980	Horseranch	3,000	23
1981	South Kechika	4,000	70
1982	South Kechika	7,000	89
1983	South Kechika	10,000	105
1984	Muskwa	7,000	182
1985	Kechika (including Horseranch)	18,000	157
1986	Muskwa	14,000	198
1987	Muskwa	10,000	125
TOTAL			996

SOURCE: Elliott 1989.

stop the decline and to evaluate the role that wolf predation played in the dynamics of these ungulates, wolves were removed in three areas and wolves and their prey were monitored in these and several other locations (table 5.2).   One wolf removal area was in the Kechika drainage in the area of the Horseranch caribou herd between 1978 and 1980 and again in 1985 (Bergerud and Elliott 1986).   The second was in the upper Kechika drainage between 1982 and 1985 (Elliott 1986a; Elliott 1989), and the third was in the Muskwa drainage between 1984 and 1987 (Elliott 1986b; Elliott 1989; figure 5.6).   To test the effects of wolf removal on ungulate densities and the proportion of young in the population, several non-treated areas were studied: Spatsizi and Level-Kawdy for caribou; and Pink Mountain, Core, and Liard for moose, and 10 census areas for sheep.   Elk recruitment rates were recorded only before and during wolf removal.   The most thoroughly documented portion of this study was the comparison of caribou in the Horseranch area (wolf removal) to the Spatsizi and Level-Kawdy areas (no wolf removal) (Bergerud and Elliott 1986).   Data from the other wolf removal areas and unmanipulated areas were never completely analyzed and published (Bergerud 1990).

Conditions Before Wolf Reduction

The autumn "total count" census of caribou in the Spatsizi control areas was about 1800 in 1976 and 2435 in 1977.   1370 were counted in 1977 in Level-Kawdy.   On Horseranch, 246 caribou were seen in 1977 (Bergerud and Elliott 1986).   In the Spatsizi and Level-Kawdy control areas, the population averaged 7.4 and 5.2% calves respectively.   The Horseranch herd consisted of 6.3% calves

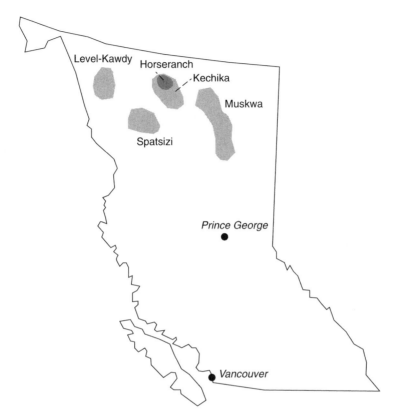

FIGURE 5.6   Location of wolf control areas in northern British Columbia, Canada.

in 1977. Wolf densities were between 9.6 and 11.2 per 1000 km^2 in all three areas before wolf reduction, however, systematic surveys of wolves were not conducted in Spatsizi or the Level-Kawdy (Bergerud and Elliott 1986).

Moose densities in 3 control areas were declining: in the Pink Mountain area they declined from 1700 ± 250/1000 km^2, to about 1100 ± 100/1000 km^2 in 5 years, in the Core area they decreased from 1000 ± 300/1000 km^2 to 800 ± 160/1000 km^2 in 3 years, and in Liard they decreased from 600 ± 80/1000 km^2 to 90 ± 30/1000 km^2 in 6 years (Elliott 1989). No single decline was statistically significant ($P > 0.05$) but all had a similar trend. The control populations contained an average of 12 calves/100 cows.

Census of the number of lambs/100 ewes and the total number of sheep seen were conducted in 10 census areas. An average of 23 lambs/100 ewes was recorded when wolves were not reduced (Elliott 1989). All sheep populations appeared to be decreasing in the early 1980s. Before wolf reductions in the

Muskwa there was an average of 16 elk calves/100 cows over 3 years (21, 14, and 13).

Wolf Reduction

Wolves in a 3000 km^2 treatment area in the Horseranch area were poisoned in 1978 and, in the following two winters, they were shot from a helicopter (Bergerud and Elliott 1986). Over 3 years, the wolf population was reduced to 38%, 11%, 11% of the pre-removal density. Wolves increased to pre-reduction levels one year after this first reduction program ended. Although not systematically censused, the wolf population in the untreated Spatsizi area was thought to have been reduced by recreational hunters who shot at least 24 individuals between 1977 and 1980 (Bergerud and Elliott 1986).

Beginning in 1982, the area of aerial wolf removal that focused on removing entire packs shifted to the upper Kechika and gradually increased in size and number of wolves removed. In 1982, 70 wolves were removed from 4000 km^2. In 1983, 89 wolves were removed from 7000 km^2, and in 1984, 105 wolves from 10,000 km^2 and finally in 1985, 157 wolves were removed from a 18,000 km^2 area that engulfed the Horseranch area again (Elliott 1989).

In 1984, the Muskwa wolf removal began with 182 wolves being removed from 7,000 km^2. In 1985, 198 were removed from an 14,000 km^2 area and in 1987, 125 were removed from a 10,000 km^2 area. Wolf censuses were conducted each year but in a variety of areas so the actual reduction in wolf densities was poorly documented (Elliott 1989; Bergerud 1990).

Response to Wolf Reduction

By the time wolf removal ended, caribou numbers in the Horseranch wolf removal area had increased from 246 to 311 (6% annual increase) (Bergerud and Elliott 1986). Two years later, caribou numbers increased further to 337. The proportion of calves in the Horseranch caribou population increased from 6.3% one year before wolf reduction to 15.2% when wolves were reduced. Over a longer period, calves/100 cows the year following wolf reduction averaged 31.8 (n=4) compared to 16.4 (n=5) when wolves were not reduced the previous winter. The total number of caribou seen during 2 censuses in each control area was about 55% of the 1977 counts, indicating decreases of 12% at Spatsizi and 11% in Level-Kawady. The proportion of calves in the untreated Spatsizi increased from an average of 7.4% in 1976 and 1977 to 10.8% while wolf control was ongoing in Horseranch. In the untreated Level-Kawdy area, the percentage calves increased from 5.2% in 1977 to an average of 13.1% between 1978 and 1983 (Bergerud and Elliott 1986).

In 1981, a radio telemetry study of caribou began in the Spatsizi control area and demonstrated that fall caribou counts were highly variable in intensity—over half the population was often not counted (Hatter 1987). Biologists debated whether or not the caribou population really decreased from the high count of 2435 in 1977 (Jones 1985). Unfortunately the Spatsizi populations were not adequately censused over the next decade. A late winter count in 1994 found 2145 caribou in alpine areas of Spatsizi, but others were known to be missed in the forest (Chichowski 1994).

While moose densities in the 3 areas from which wolves were not removed appeared to be decreasing, moose in the Middle Creek census area, which was within the Kechika wolf removal area, increased from 690 ± 130 to > 900 in 4 years (Elliott 1989). There was an average of 52 calves/100 cows over the 4 years when wolves were being removed (1982 to 1985) compared to the 12 calves/100 cows in the non-removal areas. Similarly, sheep population sizes in the 10 census areas appeared to increase after wolves were removed and the number of lambs/100 ewes increased from an average of 23 to 40. Elk calves/100 cows increased from an average of 16 when wolf numbers were not reduced to 48 when wolves were reduced.

Once active wolf reduction ended, wolf numbers increased rapidly. In the Kechika wolf removal area there were 4.6 wolves/1000 km^2 in the 16,810 km^2 area. One year later there were 12.6 wolves/1000 km^2. Because the reduction area was small, the 3-fold increase in one year was probably due largely to immigration.

The northern BC wolf reduction was more a management action than a scientific experiment, and did not have a strong experimental design. Treatment areas and wolf census areas changed from year to year and most ungulate censuses were done without the aid of telemetry. Thus no correction is possible for changes in animal distributions and sightability. However, the evidence presented suggests that the removal of 1000 wolves over 10 years (almost 800 in the last 4 years) resulted in increased populations and higher calf survival of all ungulates in the area: moose, caribou, sheep, and elk. When wolf control ended, wolf numbers increased rapidly and calf survival decreased to pre-removal levels.

Québec

Because it has 140,000 moose hunters, the economic value of moose is higher in Québec than any other jurisdiction in North America. Maintaining a high sustainable harvest is important to local culture and economies. In the 1980s, the Ministère de l'Environment et de la Faune conducted a wolf and black bear control experiment with a more structured design than most other experiments (Crête and Jolicoeur 1987). The Québec experiment, in La Vérendrey Game Reserve, used 3 areas: a wolf removal area, a black bear removal area, and an unmanipulated area with no removal. The wolf removal area was 2150 km^2, the bear removal area was 360 km^2 and the unmanipulated area was 900 km^2.

Conditions Before Wolf Control

Since the 1960s, moose density appears to have remained relatively stable at 300-400 moose/1000 km^2. A limited entry hunt kept hunting pressure low compared to adjacent areas. Wolves were studied in detail in the control area between 1980 and 1984 (Messier 1985), and density ranged from 12 to 15 individuals/ 1000 km^2. Black bears were common, at a minimum density of 230/1000 km^2.

Wolf and Bear Reduction

In the wolf removal area, between 7 and 13 wolves were removed each year by trapping or shooting from a helicopter. Indexes of wolf abundance indicated that the number of wolves in the wolf removal area decreased from 21 to 8-11 wolves/ 1000 km^2 during the control effort. In the bear removal area, 30, 26, and 25 bears were removed by trapping and shooting in 1983, 1984, and 1985 respectively (44 females and 37 males). Decreasing mean age in females captured each year (8.5, 4.8, and 3.3 for 1983, 84 and 85 respectively) and a change in trapping success (34 trap days per female in 1983, 109 trap days per female in 1985) suggested the female segment of the population was greatly reduced. The use of wolf leg-hold traps in 1983 allowed many larger male bears to escape and probably made them more difficult to trap in following years.

Response When Wolves and Bear Numbers Were Kept Low

The results of this experiment were unclear. The survival rates of calves of radio-collared cows during the control experiment were not significantly different among the 3 areas. In the unmanipulated area calf survival was 60% (n=20), in the wolf removal area it was 58% (n=24) and in the bear removal area it was 46% (n=13). The ratio of calves/100 cows was measured each year and was lowest in the unmanipulated area. From 1983 to 1986 when effects of the wolf control should have been noticed, there were an average of 24.3 calves/100 cows in the non-removal area and 39.8 in the wolf removal area. In the 3 years when bear control should have had an effect, the average of 24.3 calves/100 cows in the non-removal area compared poorly to the average of 47.3 calves/100 cows in the bear removal area. Thus, removal of neither wolves nor bears resulted in significant increases in moose calf survival. The lack of clear response was thought to be because an insufficient number of predators were removed and the fact that both wolves and bears were not removed from the same area. In addition, the experimental units may have been too small so that predators from adjacent areas hunted in the experimental area.

East-central Alaska (Fortymile, GMU 20E)

To investigate the effect of predation on the dynamics of low-density moose populations, the ADFG reduced, by aerial shooting and public harvest, wolf numbers by 28-58% for each of 3 years (1981-2 through 1983-4) in a 15,500 km^2 area in east-central Alaska on the Yukon border (Gasaway and others 1992, figure 5.7). During the 3 years of wolf control, which included both ADFG and private harvest, 58, 47, and 28% of the winter wolf population was killed each year. For 4 years after control ended, public harvest took an average of 24% of the wolf population. Black bears were scarce in the experimental area during the 1980s, but in 1984 they nonetheless accounted for 52% of moose calf mortality in the study areas, whereas only 12-15% of calf mortality was caused by wolves.

Wolf control had no measurable effect on moose calf survival. Other studies showed that predation by brown bears was the largest source of moose calf mortality and an important source of adult mortality (Boertje and others 1988). There was no evidence that nutrition, winter weather, or harvest by humans, which continued during the experiments, was limiting the moose population. Gasaway and others (1992) concluded that the combination of bear and wolf predation was limiting moose populations to a density well below habitat carrying capacity. Thus, three years of wolf control alone had no effect on this low density moose population. Therefore, although the conclusion that the population is regulated by the combined predation by wolves and bears may be correct, the experiment was insufficient to allow firm conclusions.

South-central Alaska (Nelchina, GMU 13)

Historical data concerning population changes in Nelchina wolves and caribou have been reviewed and debated by a number of authors (Van Ballenberghe 1981, 1985, 1989, 1991, Bergerud and Ballard 1988, 1989, Eberhardt and Pitcher 1992). During 1948-1954, poisoning and aerial shooting reduced wolf numbers dramatically, and it is likely that the poisoning also affected bear numbers. During and immediately after this wolf reduction period, caribou numbers increased and had more than doubled by the early 1960s. Relatively low predator numbers, favorable range conditions, relatively mild winter weather, and relatively low harvest by humans probably all contributed to this increase, but it is impossible to determine their individual effects. Upon curtailment of wolf control activities, wolf numbers also increased, likely in response to abundant prey. Over a sequence of years the combined effects of more severe winter weather and predation reduced recruitment of calves. Females were also heavily harvested by humans during this period. Again, as a result of this combination of factors, the contributions of specific factors cannot be determined, but the caribou population declined seriously.

As the caribou population declined, wolf numbers remained relatively high

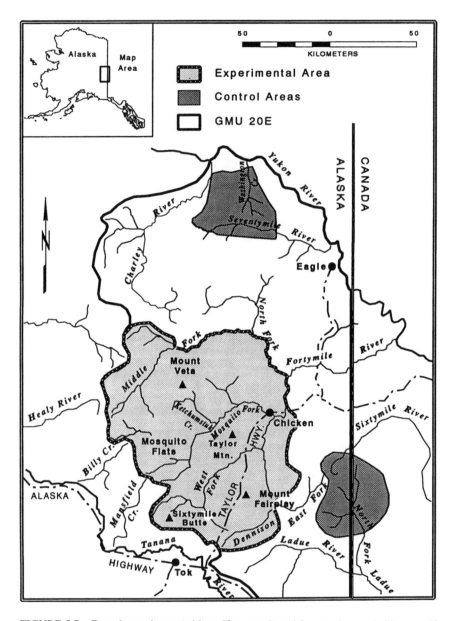

FIGURE 5.7 Experimental area (with wolf removal) and 2 control areas (without wolf removal) in GMU 20E, Alaska, and adjacent Yukon, Canada. SOURCE: Gassaway and others 1982.

and moose numbers also began to decline. Because of the high incidence of moose hair in wolf scats, wolf control was resumed on an experimental basis in 1987 and wolves were reduced by 42-58% annually (1976-1978) in a portion of the area (Ballard 1991). In this case, wolf control did not result in high annual increases in the moose population; subsequent investigations indicated that most moose calves were killed by brown bears (Ballard and others 1981).

Two other experiments were conducted to measure the effect of bear predation on moose calves when wolf densities were moderately low. First, brown bear density in a small area was lowered by 60% during May and June of 1979 by moving 47 bears to other areas (Ballard and Miller 1990). This resulted in a significant increase in the number of moose calves surviving until autumn, but because most of the bears returned to the area later that year, there were no such increases in subsequent years. Hunting regulations for brown bears were then liberalized to decrease their numbers (Miller and Ballard 1992), but no changes in calf survival could be attributed to decreased bear numbers, nor was there evidence that the bear reduction caused or contributed to the observed moose population increase.

These two experiments did not yield easily interpretable results, both because air-assisted wolf control was not associated with increased survival of moose and because the population of adult moose did not increase after bear removal. In addition, the influences of weather, predation, and habitat quality on increases and decrease of caribou populations cannot be distinguished.

GROUND-BASED WOLF CONTROL

Kenai Peninsula, Alaska

Wolves were not intentionally controlled on the Kenai Peninsula but changes associated with other human activities provide information about predator-prey relationships. During the gold rush years at the turn of the century, prospectors and others eliminated wolves from the Kenai Peninsula by hunting and trapping and poisoning them (Peterson and Woolington 1982). The prospectors also allowed fires to escape. Thus they inadvertently improved moose habitat and, at least for the short-term, deteriorated caribou habitat (Spencer and Hakala 1964). Caribou disappeared in the early 1900s. In the late 1950s wolves began to return by natural immigration (Peterson and Woolington 1982, Peterson and others 1984). During the wolves' absence, moose and bear populations had increased. In 1947, fire burned 1200 km^2 of the lowland of what was then the Kenai National Moose Range (Spencer and Hakala 1964). The burn created excellent winter habitat for 25 years and moose numbers were high until the 1970s (Oldemeyer and others 1977). Moose calf:cow ratios increased from 23:100 in 1950 to 40-50:100 in 1962. Sixty percent of the moose herd wintered in the burned area 10-12 years after the fire. Peak populations of moose in the late

1960s (more than 1600/1000 km^2) were associated with excellent forage, light hunting pressure, few wolves, and mild winters (Oldemeyer and others 1977).

By the early 1970s, the habitat in the 1947 burn was deteriorating and the moose population was showing signs of malnutrition. Cow moose were legally hunted between 1970 and 1972, there were severe winters, wolf and black bear predation on moose calves was severe (Oldemeyer and others 1977), and wolves were taking moose year round (Peterson and others 1984). The vigor of the black bear population seemed to be linked to the abundance of moose, which in turn was related to the history of forest succession. In 1967, a smaller area burned and began attracting moose in 1975. Then, when the moose population was healthy, black bears ate more moose calves and had higher cub production and survival (Schwartz and Franzmann 1991).

In the late 1970s, in an attempt to reverse a 50% decline in the moose population, private hunters and trappers were encouraged to take more wolves in a "recreational harvest" that lasted from 1976 to 1979 (Peterson and others 1984). The intensity of the harvest varied from 15% of the wolves in 1976 to 46% in 1978-79, a harvest that was similar in magnitude to air-assisted wolf control programs elsewhere. Estimates of the effects of this harvest on the wolf population were that wolf numbers were reduced following two years when the harvest exceeded 40% but that wolf numbers increased the following year when the harvest was less than 35% (Peterson and others 1984). This result is in general agreement with results elsewhere. Keith (1983) concluded that a removal of 40% of the wolves usually leads to a population decline the following year. On the Kenai, hunting and trapping reduced pack size, took many dispersing subordinate wolves, and permitted their establishment of new packs, demonstrating the extreme resilience of wolf populations to human harvest. Another major fire in 1975 began to attract moose to the burned area the following winter. Burns have a great amount of coarse woody debris and fallen trees, and are dangerous places for wolves to hunt moose. Thus moose that live in these areas are probably less vulnerable to predation and this contributes to higher moose densities in burns compared to other areas in the north. The effects of the wolf harvest on the moose population are confounded with the simultaneous improvement of habitat in the area of the burn.

Because the density of moose on the Kenai is most directly related to the history of fire, the first conclusion that can be drawn from this example is that the history of fire in an area can supersede the impact of predator control on moose. The predator control in some years in the late 1970s did reduce the wolf population and allowed more calf and old adult moose to survive than would have been the case otherwise, but the highest numbers of moose in winter occurred in areas that had been burned 10-25 years previously, regardless of the extent of reduction of wolves. A second conclusion is that a healthy moose population can support a healthy black bear population. Wolf populations were not systematically reduced, and thus, clear interpretation of the effects of increased wolf harvest in the late 1970s is not possible. The fire history was a possible confounding effect and there were no comparison sites without the treatment before and after the experiment.

Vancouver Island, British Columbia

Despite intensive forest habitat management efforts to increase numbers of black-tailed deer on Vancouver Island, deer numbers and hunter success rates began declining in the northern half of the island in the early 1970s (Archibald and others 1991). The deer declines did not appear to be related to winter severity or disease, and they occurred in both hunted and unhunted watersheds. Furthermore, deer numbers were increasing in the southern half of the island where habitat was judged to be poorer, but where wolves were less numerous. Wolf and deer studies in the late 1980s indicated that deer declines were directly correlated with wolf activity (that is, greater in the core vs. the periphery of pack territories), and that wolf predation was limiting deer population growth (Hatter 1984).

To assess the role of wolves in regulating deer populations, the British Columbia Wildlife Branch conducted an experimental wolf control program in a 2,000-km² area on northern Vancouver Island from 1982-1985. Over the 4-year period, 64 wolves were killed in an experimental area, reducing wolf density from 44 to 4-5 wolves/1,000 km². An adjacent area was designated as a non-removal zone. Measures of deer populations, including fawn:doe counts, percent juveniles in spring, and number of deer seen/km of spotlight survey route, all increased following wolf control. These indicators of population growth also increased in a comparison non-experimental area, but as many as 22 wolves were illegally killed there. Therefore, no comparative data are available from areas where wolves were not harvested. In this study, changes in hunter effort (estimated by the number of days spent hunting), a response variable of great interest to management agencies conducting wolf control, was measured. The data indicate that the decline in hunter effort would have been greater in the experimental area had there been no wolf control (Reid and Janz 1995, but see discussion of this case in chapter 5).

Because the experimental control appeared to be successful, a 3-year control program was initiated in the spring of 1986 on about 50% of the entire island, during which overall wolf density was reduced from about 13-17/1000 km² to 8-12/1000 km² (Archibald and others 1991, Reid and Janz 1995). Again, increases in fawn:doe counts and percent juveniles in spring were inversely proportional to wolf density, whereas those measures did not change in non-removal areas. Deer numbers appeared to increase during removal efforts. Data from hunters indicated that wolf control resulted in increased deer numbers. The number of hunters using the areas increased, and the number of hunter days per kill decreased in areas of wolf control. From 1983 (the year before wolf control) to 1991, the estimated resident deer hunter days in the management area where wolves were not removed declined by 73% from 12,758 to 3,618. In the adjacent management area, where wolves were removed, resident hunter days declined by only 31% from 6,359 to 4,357.

Because the 4-year period of experimental wolf removal was followed by 3 years of wolf reductions, during which an estimated 255 wolves were killed, it was not possible to assess whether the wolf reduction program had any long-term effects. Archibald and others (1991) summarized the lessons learned from the project as follows: trapping is a viable, but expensive method of wolf reduction; by controlling wolves, deer numbers will respond quickly, and when control is stopped, wolves respond quickly; and despite considerable public information and education on the value of the research projects, some residents will remain distrustful and conduct their own control programs. This case exemplifies many of the difficulties in carrying out good experimental science in the context of wildlife management.

East-central Saskatchewan

After noting a lack of response by a moose population to altered harvest regulations intended to increase herd productivity, as well as declining calf:cow ratios, biologists in east-central Saskatchewan conducted a small calf mortality study during which black bears likely killed 6 of 12 marked calves (Beaulieu 1984). Subsequently, Stewart and others (1985) removed black bears on two small areas (90 and 130 km^2) during May and June in 1983 and 1984, respectively, and found higher fall calf:cow moose ratios and percent of cows with calves on these experimental areas than on adjacent non-removal areas. For one of the areas, estimates made during the next fall indicated that both indices were nearer the original values, suggesting that the bear population had recovered to some extent. No density estimates of bears or moose were made; thus the proportion of the bear population that was removed was unknown, and no estimates of moose population densities were made. The proportion of yearlings in the population was higher after the second year, indicating a possible lingering effect of bear removal on calf survival.

NONLETHAL METHODS TO REDUCE WOLF AND BEAR PREDATION ON UNGULATES

The Alaska Department of Game and Fish has explored several nonlethal methods—diversionary feeding and sterilization of adult wolves—of regulating the impact of wolf and bear predation on moose and caribou. Fertility control was first applied to wolves in 1954 when one pair of wolves was sterilized using tubal ligation or vasectomy. Three pairs were treated in 1996 and a total of 5 pairs, one lone male, and one lone female were sterilized by 1997. Neither sample sizes nor time are sufficient to evaluate the potential effectiveness of sterilization as a method of wolf control. There have been four studies of the effectiveness of diversionary feeding on reducing moose or caribou mortality.

Diversionary Feeding of Predators

Case 1 (Boertje and others 1995)

During spring of 1985, 12 metric tons of meat (moose and scrap) were air-dropped in a 1,000 km^2 moose calving area in east-central Alaska. Brown bears, black bears, and wolves were observed feeding on the meat. The early winter calf:cow ratio in 1982-1984 varied from 26-36:100, whereas it increased to 53:100 during the early winter of 1985 immediately following diversionary feeding, a statistically significant result even though the sample sizes were relatively small (range from 17-25 observed cows). Calf:cow ratios in three untreated areas during 1985 varied between 10 and 19:100 (also significantly different from the treatment area). It was not apparent which predators were diverted. The authors also suggested that the effect of immobilization drugs on bears might have influenced their ability to forage efficiently for up to 4-5 days following immobilization.

Case 2 (Boertje and others 1995)

In the spring of 1990, 26 metric tons of moose carcasses were distributed, in equal proportions, at 61 sites during three time periods in May. Bears and wolves ate 79% of the carcasses by the middle of June. The early winter moose calf:cow ratio from 1982-1989 ranged between 12 and 38:100 cows whereas it increased to 42:100 during the early winter following diversionary feeding. Calf:cow ratios on the untreated sites during early winter 1990 varied between 11 and 27; all are significantly lower than the ratios on the experimental sites.

Case 3 (Boertje and others 1995)

In the spring of 1991, 16 metric tons of moose carcasses were distributed on the same area as in Case 2. The early winter 1991 calf:cow ratio was 32:100 on the treatment area compared with 16-37:100 on nontreatment areas. No statistical comparison was made but judging from the range of values the difference probably was not significantly different. Three alternative explanations for the result are: 1) the treatment had no effect on diverting predation or 2) the amount of supplemental food was insufficient to affect moose, particularly calf, survival rates (suggestion of the authors), or 3) too few packs were treated with supplemental feeding.

Case 4 (Valkenburg, unpublished data)

A pack of eight wolves (Nells Creek Pack) with an active den was provided with caribou and moose carcasses for a three week period during the peak calving season (mid-May to early June 1996) of the Delta caribou herd. Calves were radio-collared in 1995 to estimate their mortality rate (most of the mortality was caused by wolves from this den). Diversionary feeding resulted in a significant decrease in calf mortality during the time the caribou were near the wolf den. However, calf:cow ratios in the fall of 1995 and 1996 were not different, suggesting that 1) additional mortality factors were operating on calves once the caribou left their calving grounds or 2) predation rates increased during the rest of the summer. It is not possible to evaluate the differences between alternatives from the information presented in Valkenberg (unpublished data).

Evaluation of Diversionary Feeding Experiments

Diversionary feeding to divert wolf and bear predation on ungulates is a time-consuming and expensive process (Boertje and others 1995). In the first three examples above, most of the meat provided to wolves and bears was salvaged from auto or train accidents. Diversionary feeding during cases 1 and 2 did increase calf survival rates during the following year. Case 3 may have been unsuccessful because: 1) the amount of meat presented to predators was insufficient to change their predation rates, 2) the presentation of supplemental meat had no effect on predation rates for reasons other than the amounts of meat, 3) environmental conditions varied sufficiently between years to lessen the treatment effect, 4) or some other unknown factor.

Case 4 suggests that wolf predation on caribou calves can be reduced by diversionary feeding, but the fall calf:cow ratios were not different between treatment and nontreatment years. None of the four experiments assessed changes in prey population densities during the winter following the diversionary feeding, so it is impossible to determine whether the feeding influenced other than autumn calf:cow ratios. Also, the effects of diversionary feeding on wolf pup and bear cub survival have not been determined to assess whether diversionary feeding might actually increase predator populations.

EVALUATION OF PREDATOR CONTROL EXPERIMENTS

The results of the predator control experiments are summarized in table 5.3. Each of these control experiments was initiated because it was believed that predator control would result in increased ungulate densities and, as a consequence, increase hunter success. There was also the expectation, or at least the hope, that higher densities of both predators and prey might persist for many years after predator control stopped. Each experiment began when ungulate densities were low or declining. Hunting was sometimes simultaneously reduced. Indeed, Yukon Territory wildlife management policies stipulate that any wolf control program must be done in areas where prey are not hunted (Yukon Renewable Resources 1992).

Although considerable time and effort was expended to plan, design, and implement these experiments, the results are less informative than might have been hoped. Part of the problem is due to the size of the areas in which the experiments were carried out, the difficulty of gathering the needed information, and budgetary limits. Nevertheless, less has been learned from these experiments than would have been possible because there were serious deficiencies in their design and execution, and in the extent to which the results were monitored.

Problems in Design and Execution

Although extensive data were gathered on populations densities and trends of moose, caribou, and wolves prior to reaching a decision to initiate a control action, data on bear population densities were generally poor. In addition, assessments of habitat quality were generally limited to indirect indices of ungulate nutrition (body weight, fat deposits, bone growth, pregnancy rates, calf:cow ratios). Few direct habitat quality assessments were made. As a result, in most cases, data to support the judgments that habitats in the control area could support increased moose and caribou populations for more than a few years were not available. Several control experiments may have failed to increase ungulate populations either because predation rates by bears were high or because habitat quality was poor.

Inadequate execution makes the results of several control experiments difficult to interpret. The degree to which, and the duration over which, wolf numbers were reduced varied considerably among experiments. Sometimes wolves were reduced for only one year, and levels of reduction often fell short of what was desired. They certainly fell short of the degree that existing evidence suggests is necessary if ungulates are to increase as a result of reduced predation rates. In addition, wolf control was sometimes accompanied by other changes, such as reduction or elimination of hunting and trapping. The presence of such confounding variables makes it impossible to determine which of the factors that were altered might have caused changes in ungulate population densities.

Problems with Monitoring

The results of most experiments have been poorly monitored. In only two cases (wolf control on Vancouver Island and Finlayson) was there an attempt to measure whether hunter success—the primary goal of control—increased. Whether ungulate population densities actually increased following predator control actions was measured in only five experiments. During and after the other

TABLE 5.3 Predator Reduction Experiments Discussed in Chapter 5[a]

Method and Location	Predator Reduction		Results
	Wolves	Bears	
Air-assisted East-central AK (GMU 20A)	Wolf population reduced to 55-80% below pre-control numbers for 7 yrs (1976-1982).	Not done	Average annual rate of increase of moose populations was 15% during wolf control, and 5% for 12 yrs after the end of wolf control. Average annual rate of increase of caribou populations was 16%, and 6% for 7 yrs after the end of wolf control.
Finlayson, Yukon	49-85% of wolf population removed for 6 yrs (1983-1989); human harvest rate of moose and caribou reduced by 90%.	Not done	Increased survival of adult caribou; increased numbers of calves/100 cows for both moose and caribou. Average annual rate of increase for moose and caribou about 16-18%. Hunting success increased. Seven years after wolf control ended, moose and caribou numbers began to decline.
Southwest, Yukon	Wolf numbers reduced by 40-80% for 5 yrs (Rose Lake; 1982-1987).	Bear population reduction estimated at 7-9% (1982-1987).	No substantial increase in moose populations or cow:calf ratios during predator removal.
Aishihik, Yukon	Approximatley 76% of wolf populations removed over 4 yrs (1993-1996). Moose and caribou hunting curtailed.	Not done	Increased numbers of caribou calves/100 cows. Response of moose highly variable and not clearly related to wolf reduction. Control ended in 1996, too recently to assess long-term trends.

Location	Wolf control	Bear control	Result
Northern BC	1000 wolves removed in 10 years (1978-1987); almost 800 of which were removed in the last 4 yrs of removal.	Not done	Calf survival rates and population sizes apparently increased for all ungulates in the area. When wolf control ended, wolf numbers increased rapidly and calf survival decreased to pre-control levels.
Québec	Wolf population reduced to 48-62% over 4 years (1982-1985).	A total of 81 bears removed over a 3 yr period in a different area (1983-1985).	No apparent change in moose calf survival rate in either wolf or bear removal area.
East-central AK (GMU 20E)	Wolf population was reduced by 28-58% for 3 yrs (1981-1984).	Not done	Wolf control had no measurable effect on moose calf survival.
South-central AK (GMU 13)	Extensive aerial shooting and poisoning reduced wolf numbers dramatically (1948-1954).	Poisoning probably also reduced bear numbers.	During and after this predator reduction period, caribou numbers increased and had more than doubled by the early 1960s. This coincided with favorable weather and range conditions and low harvest by humans.
	Wolves were reduced by 42-58% for 3 yrs (1976-1978)	After wolf control ended, 60% of the bear population was trans-located or reduced by liberalized hunting regulations.	Wolf control did not result in high annual increases in the moose populations. Calf survival increased after bear removal, but bears returned to the area after summer and moose calf survival returned to levels before bear removal. No change in calf numbers could be attributed to increased bear harvests.

continued on next page

TABLE 5.3 Predator Reduction Experiments Discussed in Chapter 5[a]

Method and Location	Predator Reduction		Results
	Wolves	Bears	
Ground-based			
Kenai Peninsula, AK	Recreational harvests increased for 4 years (1976-1979).		Wolf reductions were associated with increased moose populations, but the highest numbers of moose occurred in areas that had burned 10-25 years earlier regardless of the extent of wolf reductions.
Vancouver Island, BC	Wolf density was reduced over a 4-yr period (1983-1986) to about 10% of pre-control levels.	Not done	Deer populations increased following wolf control. Hunter effort appeared to be enhanced by wolf control.
Saskatchewan	Not done	Unknown proportion of black bear population was removed in spring of 1983, and from another area in spring of 1984.	Increased calf:cow ratios after bear removal. The following year, proportion of yearlings in the moose population was higher than before predator control.

[a]See text for more details.

control experiments only indirect results, such as calf:cow ratios, were assessed, and sometimes they were determined for only one or two years. Calf:cow ratios are not always good indicators of long-term population trends and, in the short term, they can change because of changes in either calf or cow mortality rates. If cow mortality is stable, and birth rate and calf mortality both increase, then calf:cow ratios can increase with no change in the adult population size. However, most calf:cow ratios are not recorded in the late spring (shortly after parturition), but in fall to late winter. Because most calf mortality occurs in the first 6-8 months and sightability tends to be better during the late winter/early spring (pre-parturition), calf:cow ratios, or recruitment, measured at this time are a good indication of population trend and these are what have been measured.

It is apparent that political pressures have created conditions that have favored attempts to achieve quick, short-term results from predator control experiments by altering more than one factor simultaneously. In addition, budgetary constraints have led to the use of indirect measures of success, which are less expensive in the short-term but which are not good indicators of population trends. ADFG did not measure hunter behavior before and after wolf control and cannot empirically show changes in where people hunted, their success per unit effort, and their satisfaction about changes in game densities. Although in some cases prey numbers increased five fold after wolf control (20A), there is no direct empirical evidence that the social goals of predator control were achieved in Alaska.

Interpretation

What conclusions can be drawn from the predator control experiments that have been conducted in North America? Perhaps the clearest conclusion is that the experiments provide only negative evidence for the existence of an alternative stable state with relatively high numbers of both predators and prey. Only two experiments (Finlayson and GMU 20A) were monitored for long enough to reveal the existence of such a state, and the evidence from those studies was negative or equivocal. Therefore, existing evidence suggests that if predator control is to be used as a tool to increase ungulate populations, control must be both intensive and relatively frequent. There is no factual basis for the assumption that a period of intensive control for a few years can result in long-term changes in ungulate population densities.

Second, those experiments that appear to have resulted in increases in moose or caribou populations were conducted where wolves were relatively numerous, where bears were relatively uncommon or were not preying heavily on ungulate calves, where habitat quality was high, and weather was relatively benign. The evidence is not yet conclusive, but there is reason to believe that an intensive control effort, during which wolf populations are greatly reduced for several years and these other factors are favorable, can result in short-term increases in moose and caribou populations.

Third, control experiments that appeared to have had some success used methods, such as aerial shooting, that are not currently politically acceptable. Whether sufficient reductions in wolf numbers can be achieved by acceptable methods is uncertain. It is too early to evaluate the effectiveness of sterilization of wolves, and diversionary feeding appears to be feasible only in small areas and under very limited conditions. In addition, sterilization of wolves may be unacceptable to some native communities who consider this disrespectful to the wolves and, in fact, more inhumane than simply killing them.

In addition to meat collection, other problems with diversionary feeding include storage of meat over a long period of time and transportation of meat to feeding sites. Provision of commercial meat products would be extremely costly because there is no source of readily available salvage meat in Alaska. Diversionary feeding experiments can be used to evaluate the role of food abundance on predation rates in the dynamics of prey populations but they are not a feasible management tool for controlling predation over large areas.

ADAPTIVE MANAGEMENT REQUIRES AN EXPERIMENTAL APPROACH

Successful management need not be guided by a complete understanding, but enough information should be available to enable managers to make predictions and to estimate their confidence in the reliability of their predictions. To improve predictive abilities, the basic principles of experimental design can be incorporated into sample surveys and experiments. The results of properly designed experiments are more likely to be clear and unambiguous. The feasibility of experimental methods for informing policy appears to be widely underestimated (Moses and Mosteller 1995). Biologically-sound wildlife policy is most efficiently developed through adaptive management—that is, when management and research are combined so that projects are specifically designed to reveal causal relationships between interventions and outcomes. In other words, adaptive management is not simply modifying management in light of experience. It means designing management intervention to maximize what can be learned from the experiments.

Adaptive management involves (1) specification of the problem, including defining the boundaries of interest, (2) gathering all relevant information , (3) organizing it so that comparisons can be made, and (4) evaluating the strength of causal inferences that are possible (Kish 1987). In the present case, the boundaries of the current problem are populations of wolves, black bear, brown bear, moose and caribou in the natural ecosystems of interior Alaska; however, the issue is relevant to all northern ecosystems and other ecosystems where the outcome of interactions between large mammalian predators and their prey is an important wildlife management issue.

Any intentional perturbation is an experiment, but the reliability of inferences about whether the perturbation caused the observed effect depends on the design of the study. During wolf control and bear management, causal relationships have been assumed to be already known, that is, it was assumed that predators were controlling their prey. If the objective had been to test whether predators were really controlling prey, the experiments might have been planned differently.

An example of a proper design is the experiment on selected British islands to test whether populations of two species of game birds were limited by the small mammals (voles) that preyed on their nests (Marcstrom and others 1988). For five years in a row predators were removed from one area but not from a second, non-removal area. For the next four years, the removal and non-removal areas were reversed. The experiment showed that predators were, indeed, limiting game bird populations on these islands. The conclusion is strong because variation in time was controlled by doing the experiment simultaneously in two places and variation between areas was controlled by using each area for its own comparison through time. Replication, that is repetition of this design on more islands, would have yielded data on the generality of the results but would not have strengthened the conclusion.

Using the notation of Campbell and Stanley (1966), this experimental design can be represented as:

population 1   X0 X0 X0 X0 X0 0 0 0 0
population 2   0 0 0 0 0 X0 X0 X0 X0
         time →

where X represents the annual removal of predators, and 0 represents observation of the game bird population. The rows represent sets of observations through time and the columns represent simultaneous observations at a time period. It is a design leading to fairly strong causal inference because nearly all possibly confounding effects are accounted for. If the game birds had been randomly placed on different islands the design would have been a true experimental design. As it is, it has to be called a quasi-experimental design (James and McCulloch 1995). Most field experiments are in fact quasi-experiments in this sense.

Using this notation, the design of the experiment in wolf control and bear management for the Delta caribou herd in Game Management Unit 20A is:

```
population 1    0   0   0   0   XO XO XO XO 0   0   0   0   0
population 2    0                            0
population 3    0                   0                0
population 4    0                            0
population 5    0                   0                0
                time →
```

This is a multiple time series design in which the treatment (wolf removal) was applied for four consecutive years and the caribou population (0s) was monitored for a series of years both before and after the treatment in the treatment area, as well as sporadically in the four nontreatment areas. With more observations in the nontreatment areas this would be a multiple time series design, another quasi-experimental design. It allows weaker causal inference than the above design of Campbell and Stanley's study of predation on game birds, because initial differences among sites are possibly confounding. In the Delta caribou herd experiment, the results suggest that wolf removal did result in an increase in the caribou herd. The design for most of the other experiments in wolf control and bear management are either one time case studies

X 0 ,

or one group pretest-posttest designs

0 X 0,

or static group comparisons

X 0
0,
X 0 ,

all of which are inadequate for causal inference, because even though the management effort may have caused the result, the design of the experiment was inadequate for causal inference. In the first case, there was no nontreatment group and no comparison. In the second case, one group was studied before and after the treatment. This design uses the group for making the comparison, but there may have been other causes that occurred at the same time as X. In the third case there is no assurance that the initial groups did not differ. The danger is that of falsely concluding a cause and effect relationship.

None of the above designs incorporate more than one species of predator or prey. If predators switch from one prey to another, these simple designs may be able to test the hypothesis of the effects of control but modeling will be required to characterize the system satisfactorily.

To understand the degree to which predation by both wolves and brown bears limit a prey population, an experimental design like the following would be needed:

population 1	0	0	0	X	0	0	0
population 2	0	0	0	Y	0	0	0
population 3	0	0	0	XY	0	0	0
population 4	0	0	0	0	0	0	0
	time →						

where X represents wolf removal, Y represents bear removal, and 0 represents observations of the prey population.

The GMU 20A experiment does not reveal whether weather initiates population declines in general because it seems to have done just that in the Fortymile caribou herd as wolves switched from eating moose to eating caribou, and adult cows did not breed in some years with severe winters (Boertje and Gardner 1996). In the Porcupine and Denali herds calf mortality is apparently nutrition related (Whitten and others 1992; Adams and others 1995, respectively). All these ideas could be checked by studies designed by the criteria above.

In its analyses of the Delta herd (appendix C), the committee found that although calf survival was correlated with annual snowfall, incorporation of snowfall into the model did not improve its explanatory power. Modeling the dynamics of population regulation to estimate what factors are behaving in a density dependent way involves assuming a framework for how things are happening and then estimating the parameters. When based on a carefully designed and controlled experiment, one can assess cause and effect relationships. The same kind of modeling based on purely observational studies cannot distinguish between factors that regulate population densities from those that are merely correlated with them.

Prey:predator ratios (Gasaway and others 1983; Keith 1983) are more readily obtained than are full-scale area-specific studies of density, but they are subject to several major confounding variables. If prey populations are low, predators might eat more of the carcass and the prey:predator ratio will become lower. Hayes (1995) observed that wolf packs tend to kill prey at similar rates, regardless of pack size—suggesting that more of the prey carcass may be eaten when wolf packs are large. Preying on alternative prey species and prey switching may affect the results. If prey populations are close to their nutrient/climate ceiling, predation may be primarily compensatory rather than additive (Theberge 1990). Even calf:cow ratios can be misleading because they are affected by changes in mortality rates of both calves and cows. A better alternative to using ratios as response variables is to monitor the density of the population by age class and sex.

MAKING A DECISION TO INITIATE A
PREDATOR CONTROL ACTION

The committee's evaluation of the predator control experiments suggest that predator control is likely to achieve its intended goals only if a decision to initiate control is made after a thorough evaluation of the existing situation. The questions to be asked are 1) Under what circumstances is predator control justified? 2) Are there other management options than predator control available that will bring about the desired results? 3) Has an adequate analysis been made of the biologically, economically, and sociologically relevant information? and 4) Is it possible to design a control program that will include the ability to monitor the outcome and assess the degree of success?

Even if killing predators can be reasonably expected to cause an increase in ungulate numbers, it is worthwhile to consider if there are alternative, nonlethal means available to reduce the impact of predation. Under certain specific circumstances, they may be more cost-effective, and commonly, they are more socially acceptable (Boertje and others 1995). Perhaps habitat can be manipulated (burned, logged, or otherwise set back successionally) to improve its structure, quality, and distribution. A fire policy integrated with other agencies may be critical for maintaining optimal long-term habitat that will support relatively higher ungulate numbers. Also, the distribution of quality habitat, not just its quantity, may have significant positive effects on ungulate populations.

Translocation of wolves or bears is frequently only a temporary solution unless individuals are translocated to sites far enough from the study area so they do not return. However, it has been shown to be temporarily effective in some studies, and it may be applicable in special circumstances—such as in the Kenai Peninsula, where development in the Anchorage area creates a partial barrier to migration. Sterilization of wolves to offset their high reproductive rate is currently under investigation as a effective technique to keep wolf numbers at lower levels. Its effectiveness may be compromised by the tendency of wolf packs to kill at similar rates, regardless of pack size (Hayes 1995), and by the frequency with which unrelated wolves join packs, particularly in Alaska (for example, Meier and others 1995).

Diversionary feeding is likely to be more effective for bears than for wolves because the period of bear predation on ungulates is restricted to a much shorter time period. Saving calves from wolves during June is of little value if they kill calves during the remainder of the year. However, diversionary feeding of bears in combination with wolf reductions may be an effective strategy.

A decision to reduce predator numbers dramatically over a relatively short period of time will have consequences for the predators themselves. The time it takes for a species to recover to pre-control numbers or more depends not only on their inherent rates of reproduction, but also the magnitude of reduction of human-caused mortality. Wolves breed much more rapidly than bears, and thus their potential rate of increase is higher, and they can withstand higher mortality rates.

Also, the proximity of other wolf or bear populations that might provide immigrants will affect the rates of local population increase following control. In comparison to bears, wolves can more easily repopulate an area through immigration, because both males and females disperse from natal ranges. Consequently, an assessment of the geographic scale of removal efforts is important not only with regard to ungulate populations of concern, but also to the effectiveness with which the predator population can be held at necessary low levels.

Finally, wolf control may provide more food per capita for bears, but the numerical response of bears to changes in food abundance are also much lower than for wolves. Thus bears are more susceptible to control efforts, and will take much longer to rebuild their population.

An additional but certainly important concern is what the larger, unintended ecological consequences of such control will be. Even in a qualitative sense, the potential changes on other plant and animals species and systems must be considered. If ungulate numbers increase, competition with other herbivores will likely increase, and perhaps the numbers of those competitors, such as beavers, will go down. With reduced wolf and/or bear numbers, predation on alternate or less common prey species will decrease. One can imagine that in some areas, sheep numbers might increase to or above their carrying capacity, with a subsequent deterioration in habitat and a major population decline. Although this is a "what if . . .?" kind of assessment, it points out that potential unintended consequences can ultimately have large repercussions, biologically and politically.

It is clear from the experiments already done that killing a substantial proportion of the population in the target area is essential if wolf control is implemented to increase ungulate numbers. A wolf population probably should be reduced so that it is no greater than 40% of its original numbers, and then maintained there for at least 3, and probably 5 years, depending on the response of the ungulate populations. Bears, on the other hand, may need to be removed at lower rates for shorter periods of time to achieve notable results. However, since wolves have a greater per capita effect on ungulates than do bears, and bear predation behavior may be more unpredictable, results of any bear control efforts have less chance of being reliably predicted.

Another concern is which individuals in the population are going to be killed. Usually, wolves are killed by a combination of private and government efforts from the air and/or ground that targets whole packs. Focusing on specific geographic areas increases the chances that most or all of identified packs might be removed by ground-based efforts, and aerial hunting particularly if one or more pack members have been radio-marked, increases the chances of removing an entire pack (e.g., Hayes and others 1991). This is important because the removal of 60% of the members of each pack, for example, will probably have less of an effect on ungulate population change than will removal of 60% of the packs (for example, Walters and others 1981); this is particularly likely if experienced individuals are the ones left in partially affected packs. Logistical constraints likely limit the opportunities to remove entire packs, and good data comparing the consequences of different removal regimes need to be collected to identify costs and benefits of each.

Seasonal changes in predator-prey ecology are another area of concern. Reducing pack size in late autumn has little effect on winter predation rate (Hayes et al. 1991, Hayes 1995), but it should have a greater effect in summer when pack cohesion disintegrates during denning, subadults roam alone or in small groups, and many more moose or caribou calves are available than in winter. Reducing the number of wolves in summer reduces the number of independent hunting units, which in turn should reduce the predation rate of the whole population. Predation will be little changed by reducing the number of wolves in winter because the pack acts as a single unit, and kill rates of various size packs are similar (Hayes 1995).

The economic and public relations cost of predator control is high, and consequently, detailed planning is necessary to ensure a successful program. A cost effective decision process involves a series of steps beginning with the least expensive and progressing to those that are more costly.

Guidelines for Decision-making

The following is a set of guidelines, which, if followed, should increase the probability of success from implementing a management option. The committee emphasizes, however, that even if proper procedures are followed, success cannot be guaranteed, and presenting guidelines does not constitute endorsement by the committee of specific control or management activities.

The first step in deciding whether or not to reduce predators is to identify the reasons for wanting more ungulates because not all reasons are equally important. Some reasons for wanting to increase prey numbers include:

I.   Reasons for increasing ungulate populations.

A. Biological emergency. It has been suggested that predator control be considered if there is a "biological emergency"; that is, when a local ungulate population is at risk of extirpation. Local extirpations, and recolonizations have likely occurred frequently in the past, but recolonization may be less likely today in areas where people have substantially altered habitats. If it is determined that a biological emergency exists, the management agency should proceed to III.

B. Subsistence emergency. In portions of Alaska where indigenous people rely largely on ungulates, extirpation or greatly reduced populations of moose and caribou would cause significant hardship. In the past, indigenous peoples moved during periods of scarcity to where game were more abundant, but human migrations are difficult or impossible in most areas today. If predator management to maintain a satisfactory abundance of ungulates to support an indigenous subsistence lifestyle is a high priority, proceed to II.

C. Lifestyle and recreational demands. Abundant ungulate numbers are desired by people whose lifestyle includes recreational hunting. If game numbers and resulting hunting success is low, other protein source are available. Predator reduction to support higher ungulate numbers to satisfy recreational demands is currently a lower priority than for biological or subsistence emergencies.

D. Viewing and tourism demands. Abundant ungulate numbers are of value to tourists and the tourist industry. Predator reductions to stimulate an abundance of ungulates to increase tourist satisfaction is also currently a lower priority than for biological or subsistence emergencies.

II. Quantifying demand and investigative modeling. Once a compelling reason for predator reduction has been identified, the demand for an increase in ungulate populations should be quantified. Combining the use of questionnaires, personal interviews, historic hunter success records, and comparisons among areas should quantify the unmet demand and indicate the degree that ungulate numbers must increase to meet the demand. Simple population projection models and benefit/cost models should be used to determine the level and duration of predator removal that may be necessary to increase ungulate numbers to where the demand would be met, and to estimate the costs of predator removal. If, after quantifying the demand and investigating the costs of control, predator reduction might be effective, then the agency should proceed to section III.

III. Ecological investigations. Ecological investigations are expensive but are needed to assess the likelihood that a predator management activity would achieve its desired goals. The needed data include:

A. Historic trends. Are there fewer ungulates now than before, how reliable are the data, and what has changed? Could changes be explained by weather patterns, habitat changes, or human harvest?

B. Current trend. Is the population currently decreasing? This is an expensive but important step that is best estimated by monitoring adult survival using telemetry and/or a precise series of censuses.

C. Emigration. Do adjacent areas have increased ungulate populations that indicate possible emigration from the area of interest?

D. Habitat condition. Are the prey at or close to current environmental carrying capacity? Combining winter severity indices with body condition indexes such as body weight, body fat content, growth rates, pregnancy rates over a series of winters should indicate if habitat is limiting. Direct habitat monitoring should also be used, recognizing that relationships between food and cover quantity, quality, and distribution are complex.

E. Predator ecology. What are the densities of wolves and bears in the area and what are their major seasonal foods? Which predator species might be depressing ungulate numbers, and how difficult and expensive would reduction programs for each species be in the particular area?

F. Limiting factors. An expensive yet critical step is determining the relative importance of the various limiting factors.

G. Ecological consequences. Would predator control cause other negative ecological consequences, such as disrupting scavenger populations or other predator species who might depend on prey killed by wolves or bears?

IV. Management Options. If the availability of prey is significantly less than the demand, the following options should be investigated.

A. Habitat manipulation. Is it feasible to increase ungulate reproduction or decrease predation rates by improving the quantity, quality, or distribution of habitats?

B. Nonlethal methods. What is the potential for diversionary feeding, sterilization, and translocation?

C. Selective removal. Would selectively removing individual bears or individual wolf packs be effective?

D. Timing of removal. Are there certain times of the year when removal of predators would be most effective?

E. Removal methods. Assess removal methods to determine how recreational and economic benefits (hunting and trapping) might be realized while encouraging the most humane and efficient methods that are politically acceptable.

F. Removal Locations. Can predator control be concentrated in the most critical areas to maximize effectiveness while minimizing effects on the predator population?

V. Monitoring Predator Reductions. Most past predator management programs have resulted in unclear results. Control actions have sometimes been of insufficient magnitude, duration, or geographic extent to show clear results. Additionally, pre and post-treatment monitoring have sometimes been insufficient, non-experimental areas have not been maintained, and climatic conditions have often been poorly measured. Wherever possible, predator control programs should be incorporated into a reviewed experimental design to ensure that knowledge is one of the benefits of the reduction program.

REFERENCES

Adams LG, BW Dale, and LD Mech. 1995. Wolf predation on caribou calves in Denali National Park, Alaska. *In* LN Carbyn, SH Fritts, and DR Seip, Eds. Ecology and conservation of wolves in a changing world. Canadian Circumpolar Inst, Univ Alberta, Edmonton, 1995.

Archibald WR, D Janz, and K Atkinson. 1991. Wolf control: a management dilemma. Transactions of the North American Wildlife and Natural Resources Conference 56:497-511.

Ballard WB, TH Spraker, and KP Taylor. 1981. Causes of moose calf mortality in southcentral Alaska. J Wildl Manage 45:335-342.

Ballard WB. 1991. Management of predators and their prey: the Alaskan experience. Transactions of the North American Wildlife and Natural Resources Conference 56:527-538.

Ballard WB and SD Miller. 1990. Effects of reducing brown bear density on moose calf survival in south-central Alaska. Alces 26:9-13.

Beaulieu R. 1984. Moose calf mortality study. Saskatchewan Parks and Renewable Resources, Wildlife Population Management Information Base, 84-WPM-8. 5 Pp.

Bergerud AT and JP Elliott. 1986. Dynamics of caribou and wolves in northern British Columbia. Can J Zool 64:1515-1529.

Bergerud AT and WB Ballard. 1988. Wolf predation on caribou: the Nelchina herd case history, a different interpretation. J Wildl Manage 52:344-357.

Bergerud AT and WB Ballard. 1989. Wolf predation on Nelchina caribou: a reply. J Wildl Manage 42:344-357.

Bergerud AT. 1990. A review of the dynamics of ungulates and wolves in Northeastern British Columbia 1976-1990. Draft Copy. Bergerud and Associates, Fulford Harbour, BC.

Boertje RD, WC Gasaway, DV Grangaard, and DG Kellyhouse. 1988. Predation on moose and caribou by radio-collared grizzly bears in east central Alaska. Can J Zoology. 66:2492-2499.

Boertje RD, DG Kellyhouse, and RD Hayes. 1995. Methods for reducing natural predation on moose in Alaska and Yukon: an evaluation. Pp. 505-513 *in* LN Carbyn, SH Fritts, and DR Seip, Eds. Ecology and conservation of wolves in a changing world. Canadian Circumpolar Institute, Occasional Publication No. 35, Edmonton.

Boertje RD, P Valkenburg, and ME McNay. 1996. Increases in moose, caribou, and wolves following wolf control in Alaska. J Wildl Manage 60:474-489.

Boutin S. 1992. Predation and moose population dynamics: a critique. J Wildl Manage 56:116-127.

Campbell DT and JC Stanley. 1966. Experimental and quasi-experimental designs for research. Rand McNally, Chicago.

Chichowski D. 1994. Summary—BC Parks Skeena wildlife surveys, March 1994. BC Parks Branch, Smithers, BC.

Crête M and H Jolicoeur. 1987. Impact of wolf and black bear removal on cow:calf ratio and moose density in southwestern Québec. Alces 23:61-87.

Eberhardt LL and KW Pitcher. 1992. A further analysis of the Nelchina caribou and wolf data. Wildl Soc Bulletin 20:385-395.

Elliott JP. 1986a. Muskwa wolf management project of northeastern British Columbia: 1985-86 Annual Report. Wildl. Work. Rep. No. WR-21. Wildlife Branch. Ft. St. John, BC. 15 Pp.

Elliott JP. 1986b. Kechika enhancement project of northeastern British Columbia wolf/ungulate management: 1985-86 Annual Report. Wildl Work Rep No. WR-20. Wildlife Branch, Ft. St. John, BC. 17 Pp.

Elliott JP. 1989. Wolves and ungulates in British Columbia's northeast. *In* Wolf-prey dynamics and management. Proceedings. May 10-11, University of British Columbia. Vancouver, Canada. Wildl Working Report No. WR-40. BC Ministry of Environment, Victoria, BC.

Farnell R and RD Hayes. 1992. Results of wolf removal on wolves and caribou in the Finlayson study area, Yukon, 1983-92. Dep Ren Res Whitehorse, YK.

Farnell R, RD Hayes, and N Barichello. In preparation. Effects of reducing the number of wolves and human hunting on a population of caribou in the east-central Yukon from 1983-91.

Gasaway WC, RO Stephenson, JL Davis, PEK Shepherd, and OE Burris. 1983. Interrelationships of wolves, prey and man in interior Alaska. Wildlife Monographs 84:1-50.

Gasaway WC, RD Boertje, DV Grangaard, DG Kellyhouse, RO Stephenson, and DG Larsen. 1992. The role of predation in limiting moose at low densities in Alaska and Yukon and implications for conservation. Wildl Monogr 120:1-59.

Hatter DF. 1987. Perspectives on inventory of caribou in British Columbia. Spatsizi Association for Biological Research Rep No. 4. Wildlife Report No. R-14. BC Ministry of Environment, Victoria, BC. 93 Pp.

Hatter IW. 1984. Effects of wolf predation on recruitment of black-tailed deer on northeastern Vancouver Island. MS thesis, University of Idaho, Moscow. 156 Pp.

Hayes RD, AM Baer, and DG Larsen. 1991. Population dynamics and prey relationships of an exploited and recovering wolf population in the southern Yukon. Yukon Fish and Wildl Br Final Rep. TR-91-1. Whitehorse. 67 Pp.

Hayes RD. 1992. An experimental design to test wolf regulation of ungulates in the Aishihik area, southwest Yukon. Yukon Fish and Wildlife Branch. Whitehorse.

Hayes RD. 1995. Numerical and functional response of wolves, and regulation of moose in the Yukon. M.Sc. thesis., Simon Fraser University, Burnaby, British Columbia.

James FC and CE McCulloch. 1995. The strength of inferences about causes of trends in populations. Pp. 40-51 *in* TE Martin and DM Finch, Eds. Ecology and management of neotropical migratory birds: a synthesis and review of critical issues. Oxford University Press.

Jones G. 1985. Management of caribou in Spatsizi Plateau Wilderness park. *In* R Page, Ed. Caribou research and management in British Columbia. Proceedings of a workshop. BC Ministry of Forests, Research Branch, WHR-27, Victoria, BC. Southwest Yukon.

Keith LB. 1983. Population dynamics of wolves. Pp. 66-67 *in* LN Carbyn, Ed. Wolves in Canada and Alaska, Canadian Wildlife Service Report Series 45. 135 Pp.

Kish L. 1987. Statistical design for research. Wiley, New York.

Larson DG, DA Gauthier, and RL Markel. 1989a. Causes and rate of moose mortality in the southwest Yukon. J Wildl Manage 53:548-557.

Larson DG, DA Gauthier, RL Markel, and RD Hayes. 1989b. Limiting factors on moose population growth in the southwest Yukon. Yukon Fish and Wildl. Br. Rep. Whitehorse. 105 Pp.

Larsen DG and RMP Ward. 1995. Moose population characteristics in the Frances Lake and North Canol areas. 1991. Yukon Fish and Wildlife Branch PR-95-1.

Marcstrom V, RE Kenward, and E Engren. 1988. The impact of predation on boreal tetraonids during vole cycles: an experimental study. J Anim Ecol 57:859-72.

Meier TJ, LD Burch, D Mech, and L Adams. 1995. Pack structure and genetic relatedness among wolf packs in a naturally regulated population. Pp. 293-302 *in* LN Carbyn, SH Fritts, and DR Seip, Eds. Ecology and conservation of wolves in a changing world. Canadian Circumpolar Institute, Occasional Publication No. 35, Edmonton.

Messier F. 1985. Social organization, spatial distribution and population density of wolves in relation to moose density. Can J Zool 63:1068-1077.

Miller SD, and WB Ballard. 1992. Analysis of an effort to increase moose calf survivorship by increased hunting of brown bears on south-central Alaska. Wildlife Society Bulletin 20:445-454.

Moses LE and F Mosteller. 1995. Experimentation: Just do it! Manuscript.

Oldemeyer JL, AW Franzmann, AL Brundage, PD Arneson, and A Flynn. 1977. Browse quality and the Kenai moose population. J Wildl Manage 41:533-542.

Peterson RO and JD Woolington. 1982. The apparent extirpation and reappearance of wolves on the Kenai Peninsula, Alaska. Pp. 334-344 *in* FH Harrington and PC Paquet, Eds. Wolves of the world. Noyes Publ, Park Ridge, NJ.

Peterson RO, JD Woolington, and TN Bailey. 1984. Wolves on the Kenai Peninsula, Alaska. Wildl Monogr 88:1-52

Reid R and D Janz. 1995. Economic evaluation of Vancouver Island wolf control. Pp. 515-521 *in* LN Carbyn, SH Fritts, and DR Seip, Eds. Ecology and conservation of wolves in a changing world. Canadian Circumpolar Institute, Occasional Publication No. 35, Edmonton.

Schwartz CC and AW Franzmann. 1991. Interrelationship of black bears to moose and forest succession in the northern coniferous forest. Wildl Monogr 88:1-52

Spencer DL and JB Hakala. 1964. Moose and forests on the Kenai. Tall Timbers Fire Ecology Conf 3:11-33.

Stewart RR, EH Kowal, R Beaulieu, and TW Rock. 1985. The impact of black bear removal on moose calf survival in east-central Saskatchewan. Alces 21:403-418.

Theberge JB. 1990. Potentials for misinterpreting impacts of wolf predation through prey:predator ratios. Wildl Soc Bulletin 18:188-192.

Theberge JB and DA Gauthier. 1985. Models of wolf-ungulate relationships: when is wolf control justified? Wildl Soc Bulletin. 13:449-458.

Van Ballenberghe V. 1981. Population dynamics of wolves in the Nelchina basin, south-central Alaska. Pp. 1246-1258 *in* JA Chapman and D Pursley, Eds. Proceedings of the Worldwide Furbearer Conference, Worldwide Furbearer Conference, Inc., Frostburg, MD.

Van Ballenberghe V. 1985. Wolf predation on caribou: the Nelchina herd case history. J Wildl Manage 49:711-720

Van Ballenberghe V. 1989. Wolf predation on the Nelchina caribou herd: a comment. J Wildl Manage 53:243-250

Van Ballenberghe V. 1991. Forty years of wolf management in the Nelchina basin, south-central Alaska: a critical review. Transactions of the North American Wildlife and Natural Resources Conference 56:561-566.

Walters CJ, M Stocker, and GC Haber. 1981. Simulation and optimization models for a wolf-ungulate system. Pp. 317-337 *in* CW Fowler and TD Smith, Eds. Dynamics of large mammal populations. John Wiley and Sons, NY.

Whitten KR, GW Garner, FJ Mauer, and RB Harris. 1992. Productivity and early calf survival in the porcupine caribou herd. J Wildl Manage 56:201-212.

Yukon Fish and Wildlife Branch. 1994. An evaluation of calf survival in the Aishihik caribou herd, Southwestern Yukon.

Yukon Fish and Wildlife Branch. 1996. Summary of wildlife information Aishihik-Kluane caribou recovery program.

6

Social and Economic Implications of Predator Control

INTRODUCTION

In its review, in chapter 5, of biological data upon which wildlife management decisions in Alaska are based, the committee did not attempt to address questions of the social or economic implications of those decisions. Alaskan environments and the life-history traits of the focal species were described in purely scientific terms. The committee evaluated the success of past control efforts according to the stated goal of increasing moose and caribou for human harvest, without assessing whether the goal itself is economically or socially desirable. That question is addressed in this chapter, in terms of what is known about the attitudes and values of the public in Alaska and elsewhere, about the economic impacts of predator control in Alaska and elsewhere, and about how this information has been brought to bear on decision-making regarding predator control in Alaska.

There are 2 reasons for raising the issues considered in this chapter. The first is that they were specifically identified in Governor Knowles' request to the National Academy of Sciences; the statement of task directed the committee to examine what studies and research methods could be used to evaluate the full economic costs and benefits of predator control. The second is that in the committee's view economic impacts are central to modern wildlife management and depend, to a large degree, on public values. In democratic societies, government programs, such as wildlife management, should be based on widely shared public goals that legitimize the actions taken and the allocation of public financial and human resources to them. Funds for wildlife management are largely from license sales and federal aid from excise taxes on firearms and ammunition. Nevertheless, some general public funds are appropriated for wildlife management, and their expenditure means that the public implicitly agrees to forgo the benefits of other uses of the funds.

Any analysis of the social and economic dimensions of wildlife management programs must recognize that people differ in their attitudes, their values, and their economic situation. Consequently, there will be differences in how they view the programs. Accordingly, the committee organizes its discussion on the basis of the different groups that might be affected by wildlife management programs in distinctive ways—Alaska Natives, other Alaska residents, and people outside Alaska who visit it or otherwise take an interest in its wildlife. The economic and social implications of predator control programs cannot be assessed without reference to people's attitudes and values. Therefore, each case begins with an analysis of what is known about attitudes toward wolves, bears, and their primary prey followed by a consideration of the economic implications of the attitudes for predator management and control options.

The first section of this chapter considers general North American attitudes toward the gray wolf and, to a lesser extent, bears, and related views of predator management. The next section focuses on views of Alaskans and visitors to Alaska. The remaining sections analyze the economic costs and benefits associated with wolf and bear control in Alaska for non-Native residents of Alaska, nonresidents, and for Alaska Natives, and the social and economic impacts of decision-making procedures themselves.

NORTH AMERICAN ATTITUDES TO WOLVES, BEARS, AND PREDATOR MANAGEMENT

Much of North America's historical treatment of wolves, bears, and other large carnivores focused on suppressing them, as reflected most dramatically in extensive efforts to reduce and even eliminate their populations (Lopez 1978; Matthiessen 1988). The wolf was particularly targeted because of its perceived threat to livestock and human settlement and its presumed competition with people for other wildlife (Dunlap 1988). The prevailing image in colonial America of wolves and, to a lesser extent, bears is reflected in John Adams's remark of 1756: "The whole continent was one continuing dismal wilderness, the haunt of wolves and bears and more savage men. Now the forests are removed, the land covered with fields of corn, orchards bending with fruit and the magnificent habitations of rational and civilized people" (Kellert 1996:104). Widespread negative perceptions persisted well into the 20th century, as illustrated by the remarks of the historian and trapper Stanley Young; an early director of the US Biological Survey, E.A. Goldman; and the first president of the New York Zoological Society, William Hornaday: "There was sort of an unwritten law of the range that no cow man would knowingly pass by a carcass of any kind without inserting in it a goodly dose of strychnine sulfate, in the hope of killing one more wolf" (Young 1946:27). "Large predatory animals, destructive of livestock and game, no longer have a place in our advancing civilization" (Dunlap 1988:51). "Of all the wild creatures of North America, none are more despicable than wolves. There is no depth of meanness, treachery or cruelty to which they do not cheerfully descend" (Kellert 1996:105).

Those unsympathetic attitudes contributed to the extirpation of wolves and brown bears from much of their range in the 48 states and their eventual listing under the terms of the US Endangered Species Act of 1973 (Kellert 1996). Radically changing views of wildlife and large carnivores during the second half of the 20th century, a view of wolves and bears as imperiled species, and increasing knowledge of these species all contributed to more-positive attitudes toward these animals. This change has been especially notable among urban, and younger Americans, and those with more extensive formal education (Kellert 1991, 1994).

A national study conducted in 1980 showed Americans roughly equally divided in positive and negative views of wolves and bears (Kellert 1985, 1980). More recent studies in Michigan, Minnesota, Idaho, Montana, and New Mexico and among visitors to various protected areas show widespread appreciation and concern for wolves and bears among a majority of respondents (Kellert 1996). A substantial minority express strong protectionist views toward the species. The wolf, in particular, has emerged as a powerful symbol for protecting and restoring wildlife and wilderness (Kellert 1996; Naess and Mysterud 1987). Popular books, films, and other media have celebrated the positive attributes and qualities of wolves and bears. Many new and established nongovernment organizations have marshaled considerable resources and public support for protecting and restoring wolf and bear populations. Studies among visitors to Yellowstone and Glacier National Parks have strongly endorsed the existence value of those species and supported the reintroduction and augmentation of populations in these areas (Bath 1987, 1991; Biggs 1988; Jobes 1991; McNaught 1987; Reading and others 1994).

Considerable ambivalence remains, however, among most demographic groups regarding wolves and, to a lesser extent, bears (Arthur and others 1977; Bath 1991; Buys 1975; Herrero 1970; Hook and Robinson 1982; Jonkel 1975; Kellert 1996; McNamee 1984; Shepard and Sanders 1985). Largely hostile and exploitative attitudes occur among resource-dependent and rural groups, especially livestock producers (Herrero 1978; Kellert 1986, 1994, 1996; Pelton and others 1976; Stuby and others 1979). Negative, albeit less-antagonistic views have also been observed among the elderly and people with less formal education. Recreational hunters and commercial trappers often reveal positive attitudes toward and greater knowledge about these species than most groups, but are inclined to support their harvest and control. Various studies suggest that positive sentiments toward wolves and bears typically increase with geographic distance from either extant or proposed reintroduced populations (Llewellyn 1978).

Attitudes toward and knowledge of wolves and bears are often surprisingly independent (Burghardt and others 1972; Kellert and others 1996; Murray 1975; Peek and others 1991; Petko-Seus 1985). Both people who are strongly in favor and those who are strongly opposed to protection and restoration often possess relatively high levels of knowledge of the species, but most people have limited factual understanding of either animal.

North American attitudes toward controlling wolf, bear, and other large-predator populations—especially to minimize livestock depredations—have been examined (Arthur and others 1977; Braithwaite and McCool 1988; Brown and others 1981; Buys 1975; Colorado Division of Wildlife 1989; Hastings 1986; Herrero 1978; Jope and Shelby 1984; Kellert 1985; McCool and others 1990; Petko-Seus and Pelton 1984; Trahan 1987). The majority of the general public favors limiting control to situations where substantial depredations have been clearly and convincingly demonstrated. Issues of culpability and humaneness (with respect to perceived pain and suffering) tend to be more important considerations than the cost of control. Much of the general public disapproves of indiscriminate population reductions that are independent of an individual wolf's or bear's guilt. They also strongly oppose using poisons, denning, or aerial gunning. Livestock producers and other resource-dependent groups are far more accepting of these latter control methods.

A number of economic implications relevant to wolf and bear management in Alaska can be drawn from those findings. Considerable potential consumer demand appears to exist among many groups of North Americans for visits to areas that have populations of wolves, bears, and other large carnivores. Results from the Boundary Waters Canoe Area, Glacier and Yellowstone national parks, and other protected areas suggest that the presence of those species constitutes a major attraction. Intensive wolf or bear management in Alaska conflicts with this potential consumer demand; additional data on Alaska tourism presented in the next section confirm this view.

Widespread public support for wolf protection and restoration in the lower 48 states and a general perception among the public of wolves and brown bears as imperiled species are likely to encourage advocacy groups to oppose wolf and bear control in Alaska. That could produce economic actions, such as support of a tourism boycott. A majority of the general public supports geographically limited predator control when a clear and compelling need has been demonstrated, when indigenous populations suffer from the absence of control, and if humane control methods are used. The need for convincing scientific data to support control programs and the use of animal-specific and humane control methods are both likely to increase management costs.

ALASKAN ATTITUDES TOWARD WILDLIFE

The Alaskan population is now largely urban: 67% of Alaskans live in cities. About 16% of the population is Alaska Native, the majority of whom reside in rural Alaska. These Alaska Natives are economically and culturally dependent on subsistence hunting. Immigrants from other states amounted to an average of roughly 10% of Alaska's population each year during the decade 1980-1990; these immigrants tend to be much younger and to have more formal education than long-term residents of Alaska.

Various surveys that have assessed attitudes of the Alaskan people are discussed below. These surveys were conducted by different groups, including the Alaska Department of Fish and Game (ADFG), other government agencies, trade organizations, and citizen interest groups. They varied considerably in sampling techniques, survey methods, and data-collection procedures, so their quality and representativeness vary. Most relied on telephone data-collection procedures, which often underrepresent rural, especially Alaskan Native, population groups. All the studies were cross-sectional (conducted at one point in time) and thus reflect bias associated with the time of the year when the studies were conducted and the particular issues and events occurring then. Nevertheless, the results collectively present a crude approximation of the views and perceptions of the Alaska public and provide a useful perspective of prevailing wildlife attitudes and values in Alaska.

According to the 1991 National Survey of Fishing, Hunting, and Wildlife-Associated Recreation, 41.2% of the Alaskan population age 16 and older participated in hunting or fishing; this is a higher percentage than that of any other state; and much higher than the national average of 21%. Some 20% of Alaskans purchase a hunting license in an average year, but the percentage has been slightly and steadily declining for more than a decade. A majority of Alaskans report that they have at some point purchased a hunting license. Less than 10% of Alaskans oppose hunting. The proportions of Alaskans who have ever hunted and who object to the activity are, respectively, considerably higher and lower than in the contiguous 48 states. In 1991, about 62.1% of the adult Alaskan population participated in some form of primary nonconsumptive wildlife activity (viewing, photographing, or feeding fish or wildlife); this is substantially higher than that in most other states and much higher than the national average of about 40%.

The harvesting of furbearers (for example, more than 1,000 wolves each year) and the amount of subsistence hunting are far greater in Alaska than elsewhere in the United States (ADFG 1994; ADFG 1994-95; Miller and McCollum 1994a). Much of the latter activity is associated with the relatively large Alaska Native population.

The nonconsumptive value of wolves and bears is suggested by both statistical and anecdotal findings among residents and visitors to Alaska. According to one study, for example, the most popular and economically important species for viewing purposes were bears, wolves, and marine mammals, with a higher preference for viewing bears than wolves (Miller and others 1997). Alaska is the only place in North America where opportunities for viewing brown and black bears are readily available.

Some 96% of Alaskans believe that wildlife adds considerably to their "enjoyment of living in Alaska" (Miller and McCollum 1994b). A majority of the Alaskan population report a desire for more state-managed areas for wildlife-viewing purposes, although only a minority of resident hunters support creating more wildlife-observation areas. Nearly three-fourths of resident hunters object to establishing such areas if it will result in closing areas to hunting. Tolerance of large carnivores among the Alaskan public is suggested by the fact that nearly half approve of bears living near urban areas, although only a minority of hunters support this idea (Miller and McCollum 1994a).

Most Alaskans endorse the hunting of bears, although a large majority object to baiting and attracting bears with food, and a majority disapprove of trophy hunting (the main objective of most bear hunting). Recreational wolf hunting generates little interest among Alaskan residents or nonresident visitors (Miller and McCollum 1994a), but wolves are extensively harvested for their pelts, and most Alaskans support this use. As with bear hunting, issues of "fair chase" and humane harvesting of wolves are important among the Alaska public. Most Alaskans object to using airplanes, denning, and poisons as methods of harvesting wolves, as discussed later (Alaska Wildlife Alliance 1992; Kellert 1980; Miller and McCollum 1994a,b,c).

Surveys of public attitudes toward controlling wolves and, to a lesser extent, bears to increase moose and caribou populations generate varied responses, depending on the wording of the questions and the group surveyed (Alaska Wildlife Alliance 1992; Miller and McCollum 1994a). In one study, a near majority (47%) of the Alaska public supported and 37% disapproved of killing wolves in some areas to increase moose and caribou populations. A majority of Alaska nonconsumptive wildlife users and female residents object to killing wolves to increase ungulate populations, but more than 60% of resident Alaskan hunters support this notion.

Strong support for wolf control has been found in areas that have experienced substantial decreases in moose or caribou populations presumably because of predators (Anderson 1995, Wolfe 1991; Gardner and Taylor 1992; Wolfe and Walker 1987). Some 90% of residents in game management unit 19D east, for example, support a reduction in wolf populations to increase moose numbers; more than 70% favor decreasing wolf populations for more than 5 years to maintain high moose numbers. A slight majority of residents of this area support brown bear control to increase moose numbers. Those views, as well as more ethnographic and anecdotal data from other areas of the state, indicate that a majority of rural, native, and especially subsistence groups support wolf control and harvest. The food, economic, and cultural importance of subsistence hunting and trapping is generally associated with support among Native populations for limiting wolf and, to a lesser extent, bear populations in areas where ungulates have declined presumably because of predators.

A majority of nonindigenous Alaskans, particularly residents of urban areas and females, object to reducing wolf numbers (Alaska Wildlife Alliance 1992; Miller and McCollum 1994a). Those groups reluctantly endorse wolf control only in specific areas where clear and convincing data reveal substantial declines in moose and caribou numbers because of wolf predation. Moreover, one survey found that 43% of Alaskans believe that the overall number of wolves killed for control purposes should be reduced, whereas only 8% favored increasing the number of wolves killed for this reason. A majority of Alaskans remain unconvinced that reducing wolf numbers will increase moose and caribou hunting opportunities.

Public attitudes toward different methods of wolf and bear control have also been examined (Alaska Wildlife Alliance 1992; Kellert 1980; McCollum and others 1996; Miller and McCollum 1994a). Most data reveal an aversion to aerial techniques, as the outcome of a recent statewide referendum corroborates. Results from Alaska, as well as other areas of the United States, suggest that issues of fair chase and humaneness are major determinants of public opinion and often override economic considerations. For example, most people oppose poisoning, denning, and aerial gunning even when these techniques appear cost-effective. Far greater support exists for ground-based hunting, trapping, and snaring to kill wolves, although majorities of female, urban, and antihunting groups object to these control methods as well. Most Alaskans support wolf and bear control by local hunters rather than by professional wildlife personnel. Many, especially rural and native groups, believe that wolf and bear control should generate practical gains, such as meat and furs.

The views of key Alaskan groups suggest the following conclusions. Majorities of urban Alaskans, nonconsumptive users, younger people, females, and people with higher levels of formal education tend to object to wolf control and, especially, bear control unless a substantial problem of declining moose and caribou populations has been clearly and convincingly demonstrated in geographically specific areas. Those groups often object to control methods perceived as inhumane (for example, likely to cause pain and suffering) or unfair (for example, violating presumptions of a fair chase). The views of Alaska sport hunters tend to be more diverse. Strong support exists among resident sport hunters for wolf control and sometimes bear control in areas where substantial declines in large ungulate populations have occurred and that afford relatively easy road access for hunting purposes. Most Alaska sport hunters object to indiscriminate wolf reduction throughout the state and to control methods regarded as unsportsmanlike (for example, airborne hunting). Subsistence hunters, especially rural and native peoples, generally support wolf and bear control for cultural, economic, and food-gathering benefits.

For 2 reasons, the cost of a wolf and bear control effort is likely to increase substantially if prevailing attitudes among the Alaskan general public, especially urban residents and nonhunters, persist. First, most of the public objects to wolf and bear control except in specific geographic areas where a serious decrease in ungulate populations has been clearly and convincingly related to wolf or bear predation and where serious economic or cultural harm has been inflicted on resident sport or subsistence hunters; demonstrating and communicating such a magnitude of effect will inevitably increase management costs. Second, most of the Alaskan public objects to control methods perceived as inhumane or unfair—for example, aerial control, denning, toxicants, and trophy hunting. Other control methods tend to be more costly, labor-intensive, time-consuming, or technically challenging. A partially offsetting economic effect might be associated with widespread support for wolf control by local people who derive economic and practical benefits from the activity.

Alaska has experienced a slight but consistent decline in the number of licensed hunters during the last decade and more. It might be due to decreasing ungulate numbers in areas readily accessible to large urban populations. But absolute and proportionate declines in the number of licensed hunters have been observed throughout the United States. If the decline in ungulate numbers in areas readily accessible to hunters has contributed to a decline in hunting participation, wolf and bear control could produce positive economic benefits. Brown bear control could also generate economic returns if conducted by nonresident trophy hunters, although widespread opposition to trophy hunting suggests that the meat and possibly pelts should be distributed to local populations. A greater harvest of ungulates resulting from wolf and bear control would produce substantial economic and cultural benefits for subsistence and native populations in rural Alaska.

Most nonconsumptive wildlife users and urban wildlife enthusiasts believe that intensive wolf and bear control reflects an entrenched bias among Alaska game management officials and hunting interests. That view encourages the use of legislative and judiciary methods for promoting policy change, rather than attempts to exercise influence on administrative agencies. The success of the 1997 ballot initiative involving aerial hunting of predators will probably reinforce the perception, which is likely to result in substantially greater management-associated costs because of the heavy economic burden created by using legislation and litigation as a means of achieving policy goals. Many nonconsumptive wildlife users also view ADFG as less interested in providing recreational viewing than hunting opportunities; this perception is likely to encourage opposition to user fees and taxes on nonconsumptive users as a way of generating additional wildlife agency revenues.

CONCEPTUAL FRAMEWORK FOR ECONOMIC ASSESSMENT

From an economic perspective, Alaska has at least 3 very striking features. First, it is a huge area—more than one-half million square miles—with a population of about 610,000, most of whom reside in and around 3 cities (Alaska Department of Labor 1996). Second, about 60% of the land is in federal ownership, 28% is in state ownership, and 12% is Native lands (figure 6.1). Only 1% is private, non-Native land. Third, natural resources and resource-based activities, such as petroleum extraction, tourism, and hunting, play a dominant role in the state's economy. The petroleum industry accounted for more than one-third of the gross state product (GSP) in 1993. More than one-fourth of all employment in the state is government-related, including federal, state, and local government employees. The seafood industry is the state's largest private employer and the tourist industry is the second-largest private sector employer, annually attracting nearly 1 million visitors who spend more than $1 billion of a total GSP of about $24 billion. Alaskans themselves are estimated to have spent about $89 million on hunting, $239 million on fishing, and $144 million on primary nonconsumptive wildlife activity in 1991.

Although hunting and tourism provide an important contribution to the Alaskan economy, there are 2 reasons why the data on expenditure and employment can be misleading in assessing the economic benefits of wolf control or the economic consequences of a tourism disruption triggered by public opposition to wolf control. The first reason has to do with the distinction between total and marginal economic impacts on the economy. The total impact is the entire contribution of a sector. The marginal impact is the increase or decrease in the contribution associated with some expansion or contraction in the activity within the sector. If what is at issue is the total elimination of a sector, whether hunting or tourism, the total impact is the appropriate measure to use. But if the change is smaller than that, the correct measure is the marginal impact of the change. For example, if it involves some increase in hunting or some decrease in tourism, the appropriate quantity is not the total value of all hunting or all tourism, but the value of the particular increase or decrease expected to occur. The marginal impact is harder to determine because it involves forecasting the magnitude of the specific increase or decrease in activity, as well as assessing the economic value associated with it. Nevertheless, this is the appropriate approach for evaluating the economic implications of predator control policies.

The second reason is that expenditure is generally not a correct measure of value, and the increase or decrease in expenditure is generally not an appropriate monetary measure of the increase or decrease in well-being. There is a basic distinction between financial or accounting revenues or costs and economic benefits or costs. Accounting revenues and costs reflect an assessment of monetary inflows and outflows according to the principles and conventions of accounting rules. Economic costs and benefits are intended to reflect a broader social perspective based on real costs, in the sense of resources actually used up, and real benefits, in the sense of actual changes in people's well-being. Although not unrelated to financial revenues and costs, economic benefits and costs represent a different standpoint for assessment.

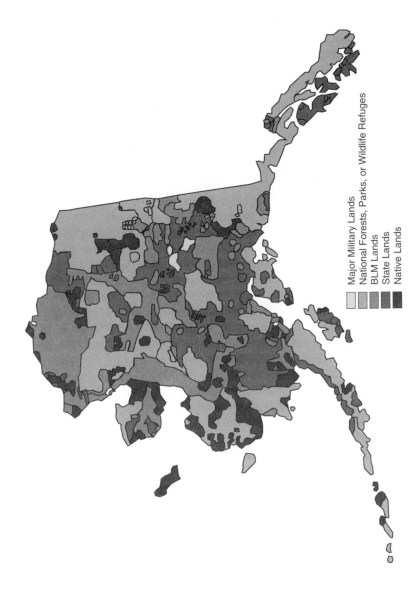

FIGURE 6.1   Distribution of public land ownership in Alaska.

Marginal Costs and Opportunity Costs

The correct concept to use in an economic assessment of the costs of wolf control is the economic concept of *marginal cost* (Freeman 1993, Turner and others 1993). This should not be confused with average cost. The average cost is simply the total cost for some item divided by the quantity of the item; marginal cost is the change in cost per unit change in the quantity of the item. If all units of a commodity cost exactly the same, there is no difference between average and marginal cost; if not, there is a difference. Such differences can arise for several reasons, including the presence of fixed costs. By definition, where there are only fixed costs, the marginal cost is zero—since cost does not vary with the quantity of the item, there is zero increment in cost when the quantity is raised, and zero decrement when it is lowered. But, although the marginal cost is zero where there are fixed costs, the average cost is positive. It can make a substantial difference, therefore, whether one uses average or marginal cost. Marginal cost is generally the relevant concept for an economic analysis because it reflects the marginal impact; for the reasons given above, this is what the analysis should aim to measure. For example, if a hunter already owns some hunting equipment such as a rifle, the marginal cost of an extra hunting trip is the ammunition and the cost of gas needed to make the trip but not the cost of the rifle. Conversely, if somebody who did not previously hunt or own a rifle is now induced to take up hunting by an improvement in the moose population, say, then the cost of a rifle counts as part of the marginal cost of the increase in hunting associated with this person.

The economic concept of marginal cost also includes costs that are implicit rather than explicit. These are known as *opportunity costs* and they reflect the notion that the real cost of an item might be not what one pays directly to obtain it, but rather what one forgoes to obtain it. For example, in the case of a government agency with a fixed budget and a fixed staff, part of the cost of providing a new activity—say, wolf control—might be the other services that have to be reduced or postponed when agency personnel are diverted to work on the new activity. There could be no direct increase in payroll, because no extra staff are hired, but substantial opportunity cost in terms of something else forgone. Another example of opportunity cost is the time spent by a hunter in traveling to and from a site and hunting there. Depending on the situation, he might otherwise have spent his time in earning income at work or doing something else of value; if so, the opportunity cost of his time counts as part of the marginal cost of the hunting trip from an economic perspective.

Willingness to Pay and Willingness to Accept

Similarly, with benefits the correct concept to use is the marginal benefit, that is, the equivalent monetary value of the change in people's well-being as a result of the program (Freeman 1993, Turner and others 1993). With marginal benefit, as with marginal cost, there are several important points of difference between economic and accounting concepts. A crucial difference is that, in economics a person's expenditure on an item is generally not a measure of the item's value to him. To be sure, there are some circumstances where expenditure is a measure of value. For small changes in marketed items that are freely divisible in quantity, the price a person pays for the last unit consumed is equated to the unit's marginal value. But for items that are not marketed or not freely divisible and for large changes in items that are marketed, the amount paid provides a bound on the value to the consumer but is not the same as this value. In economics, the difference between what an item is worth to a consumer and what he actually pays for it is known as the *consumer's surplus* from the item. The change in consumer's surplus is the economic measure of marginal benefit— the lost consumer's surplus, which is the economic measure of loss if the item is taken away, or the extra consumer's surplus, which is the gain if more of the item is provided.

In economics, there are 2 ways to express in monetary terms what an item is worth to a person, that is, 2 ways to define consumer's surplus: the compensating variation and the equivalent variation—more colloquially, willingness to pay (WTP) and willingness to accept (WTA). The WTP value of an item is defined as the maximal amount that a person would be willing to pay to ensure his continued access to the item over and above what he does pay for it; the WTA value is the minimal amount of money that he would be willing to accept to forego the item, above and beyond what he does pay for access to it. In general, the 2 measures are not equal; how close or far apart they are depends partly on whether things that money can buy are good substitutes for the item in question (Hanemann 1994).

All monetary measures of value in economics can be shown to be either WTP or WTA measures. The logic of the economic measure of value is that it expresses value in monetary terms through a determination of equivalence. Regardless of whether the item itself is a market good or can be obtained by spending money, its monetary value is defined to be the change in monetary income that would have an equivalent effect on the person's well-being. Therefore, changes in income are the currency in which economic welfare is measured.

How does this relate to expenditure? Suppose a person is able to go hunting at a cost of $150 per trip and elects to take 3 hunting trips per year. (This example assumes that the hunter can take as many trips as he wants at this price, with no constraint such as when there is only a limited number of permits for a particular type of hunting trip.) Thus, he spends $450 annually for hunting trips. However, he might be willing to pay something more than this if it were necessary. To make 3 trips, say, he might be willing to pay $500; his WTP measure of consumer's surplus would be $50. Or he might be willing to pay $1,000; his WTP measure of consumer's surplus would then be $550. In general, there is no specific relationship between the amount of expenditure and the magnitude of the consumer's surplus. However, the amount of consumer's surplus on the last unit consumed is likely to be small; the last trip is likely to be worth something close to $150. It could hardly be worth less, because if it were, he presumably would not have elected to take it. It is unlikely to be worth substantially more, because if it were, he might have taken more trips than this. Hence, the value of the last unit to the consumer is likely to be close to its cost, and the consumer's surplus is likely to be small. But for units other than the last, the value can be considerably larger than the cost, and the consumer's surplus might be substantial. All that can be said in general is that if the cancellation of a predator control program makes it less attractive for a person to go hunting, the loss of consumer's surplus should be larger when the reduction in hunting activity is greater. But the loss bears no specific relationship to the expenditure on the lost trips; rather, it captures the hunter's loss of utility in addition to the expenditure foregone. (The discussion here deals with just the *definition* of the concept of monetary value in economics rather than its measurement, which is taken up in the following section.)

How is all this related to jobs and employment? Lay discussions of economic effects are often framed in terms of jobs gained or lost. Moreover, there is a type of economic analysis that is often performed in these circumstances to determine direct and indirect effects on employment, known as economic-impact analysis or, more technically, input-output analysis. This type of analysis can be reconciled with the economic concept of value discussed above up to a point. It should be noted that most discussions of employment focus on total employment, as opposed to the marginal increase or decrease in employment that would be associated with implementing or canceling a predator control program. For the reasons given above, it is the marginal impact on employment that is appropriate in an economic assessment. Furthermore, framing the discussion in terms of jobs versus personal income from employment and other sources involves mainly a difference in choice of units, not a substantive difference in the underlying item that is being measured. In addition, discussions of jobs are often confounded by assumptions about whether the workers would otherwise be employed elsewhere and whether the same needs would ultimately be fulfilled elsewhere.

For example, suppose that an analysis reveals that because of the abandonment of a wolf control program and the subsequent reduction in hunting activity, 1,500 jobs for hunting guides and outfitters would be lost in Alaska. Before one can reach any definitive conclusion, it must be determined whether any of those workers can be expected to obtain other jobs, as opposed to being permanently unemployed. To the extent that they do obtain other jobs, the relevant measure of loss is the net loss of personal income from employment in the new jobs versus the old, not their gross income in the old jobs. A similar issue arises with respect to nonresident hunters and visitors to Alaska. To the extent that they reduce their visits or hunting trips in Alaska as the result of a government program, from a national perspective the relevant measure of loss is the net loss of enjoyment they receive as consumers after allowing for the other activities that they engage in elsewhere and the net loss of income from employment as a result of the reduction in their visits to Alaska after allowing for the income from employment generated by their increased activity elsewhere.

In summary, an economic assessment of a predator control program is couched in terms of the net gain or loss in personal income for the various parties affected by the program, allowing for the effect of employment and business profits on personal income, plus the net gain or loss in utility from the consequences of the program expressed in monetary terms as the change in income that is equivalent to this change in utility.

Assume that a predator control program does indeed trigger an increase in moose and caribou populations. If one thinks in terms of a balance sheet of the potential effects of a predator control program, there would be 3 main items: the marginal benefit of whatever increase occurs in the populations of prey species, such as moose and caribou; the marginal cost of planning and implementing the predator control program itself; and any marginal reduction in benefit associated with the decrease in the predator population. The benefits can be divided into five components:

1. The gain in utility for residents from increased recreation associated with moose and caribou.

2. The gain in utility for nonresidents from increased recreation associated with moose and caribou.

3. Any other gain in utility for residents arising from the increase in moose and caribou populations.

4. Any other gain in utility for nonresidents arising from the increase in moose and caribou populations.

5. The gain in personal income from employment and profits for residents resulting from increased recreation or tourism in Alaska by residents or nonresidents.

The costs similarly can be sub-divided into:

6. The costs to government agencies and others for planning and implementing the predator control program.

7. Any loss of utility for residents associated with the reduction in predator populations.

8. Any loss of utility for nonresidents associated with the reduction in predator populations.

9. Any loss in personal income from employment and profits for residents resulting from a reduction in tourism in Alaska by nonresidents triggered by their reaction to the predator control program.

Those benefits and costs are discussed further in the following 2 sections.

ECONOMIC IMPACTS OF PREDATOR CONTROL ON NON-NATIVE RESIDENTS AND NONRESIDENTS

Because of their cultural and economic differences, the committee treats non-Natives and nonsubsistence residents separately from Native and subsistence peoples, and nonresidents separately from residents. This section presents an economic assessment of the impact on non-Native residents and nonresidents; the next section deals with Native and subsistence peoples. The first part of this section covers the impact of a predator control program on recreation associated with the prey species, moose and caribou. The succeeding parts deal with the impact on recreation associated with the predator species, wolf and bear; non-recreation impacts, including what are known as nonuse or passive-use values; the impact on resident non-Native Alaskans' personal income from employment and business profit; and the costs of predator control programs. In each case, we consider both what economic information is available and how it appears to have been used by ADFG in its decision-making process.

Gain from Increased Recreation by Non-Native Residents Involving Prey Species

With respect to recreation by non-Native residents, the main activities likely to be affected by predator control programs are hunting and nonconsumptive wildlife-associated recreation (such as wildlife viewing and photographing). In each case, the increase in consumer's surplus associated with enhanced recreation is composed of a gain from an increase in recreation activity and a gain from an increase in enjoyment of the recreation activity. For example, if there are better hunting opportunities for moose, more people might take up hunting in Alaska and existing hunters might take more trips, go to different locations where conditions have improved, and obtain greater enjoyment from hunting at the locations where they normally hunt. Thus, there are changes both in the number of hunting trips and in the consumer's surplus per hunting trip.

Both those types of changes and the resulting change in aggregate consumer's surplus can in principle be measured by estimating a behavioral model of people's participation in hunting and choice among hunting sites, known in economics as a travel cost model. The model is so called not because it generates economic values on the basis of travel costs but rather because it is used to analyze hunter choices of how often to go hunting and where to hunt on the basis of factors likely to affect those choices, including the costs of hunting at various locations and the quality of the hunting experience there. By analyzing the implicit tradeoff between site choice and travel cost, it is possible to infer the monetary value of the hunting experience in terms of the WTP or WTA measures of consumer's surplus. This type of analysis is known in economics as revealed preference analysis—when one makes particular assumptions, people's behavior reveals their preferences. In this context, travel costs are important not because expenditures are a measure of value, but because costs shed light on the tradeoffs that people face when they make recreational choices; it is their choices from which estimates of value are inferred. The use of travel costs for this purpose was first suggested by Hotelling (1931); the first to implement this approach were Trice and Wood (1958) and Clawson and Knetsch (1966). Behavioral models of recreation choices based on travel cost are now a standard component in environmental valuation (see McConnell and others 1990 or Freeman 1993).

A travel cost model of hunting or wildlife-viewing recreation could be used to predict both the change in hunter behavior (total hunting activity and allocation of the activity among alternative sites) and the change in consumer's surplus as the result of a predator control program that causes changes in the hunting season or the abundance of animals at various locations. For sportfishing in south-central and southeastern Alaska, models like this were developed for ADFG by Carson and others (1990). Such models apparently have not yet been developed for hunting or wildlife viewing in Alaska.

In fact, it is not clear what type of economic analysis was performed by ADFG in connection with its wolf control programs in 1993 and 1995. Nevertheless, there is an active economics research program in the ADFG Division of Wildlife Conservation, and staff economists have collected data on the economics of wildlife-based recreation and prepared estimates of the consumer's surplus from hunting and wildlife viewing in Alaska (McCollum and Miller 1994). This has been applied to an economic assessment of game management strategies by Miller and others (1997).

We first summarize the findings from those studies and then comment on their applicability to the problem at hand. The estimates of consumer's surplus per trip are shown in table 6.1, which is based on material supplied to the committee by ADFG. The estimates were derived from 2 1992 surveys conducted by ADFG about wildlife in Alaska. One survey was administered to a large sample of resident voters (intended as a proxy for the adult resident population) and focused on their attitudes to wildlife and their involvement in nonconsumptive wildlife use in Alaska in 1991, especially wildlife-viewing trips. The other survey was administered to a large sample of resident and nonresident hunters who had taken out Alaskan hunting licenses in 1991; it focused on their attitudes to wildlife and their involvement in consumptive wildlife use in Alaska, especially big-game and waterfowl hunting. Both surveys selected a specific trip by the respondent and asked whether the respondent felt that the trip had been worth the cost (in the sense that they would take it again). If the answer was yes, the respondent was asked how much more they would have been willing to pay for the trip. That question uses an open-ended format, also known as a continuous-response format. An alternative approach presents respondents with a specific cost increment and asks whether they would have been willing to take the trip at

TABLE 6.1   Estimated Average Net Economic Value of Trip for Different Species

Purpose of Trip[a]	Primarily for Wildlife Viewing	Secondarily for Wildlife Viewing	Hunting	Hunting
Respondents	Alaskan Residents	Alaskan Residents[b]	Alaskan Residents	Non-Residents
Species Viewed or Hunted				
All species	$134	$ 93	$167	$440
Any bear	$170	$121	$172	$521
Black bears	$182	$140	$152	$366
Brown bears	$226	$ 15	$208	$606
Moose	$123	$ 87	$181	$393
Caribou	$135	$103	$168	$432
Dall sheep	$132	$110	$267	$492
Deer/elk	$141	$ 97	$143	$222
Mountain goat	$206	$123	$126	$419
Wolf	$195	$ 44	$1500	$351

SOURCES: DW McCollum and SM Miller 1994; DW McCollum and SM Miller, unpublished data.

[a]These results are based on a survey in which respondents indicated the primary and secondary purposes of their trips and which species they saw or hunted. For example, while hunting was the primary purpose of some trips, a secondary purpose was to watch wildlife. Note that hunting a particular species only means that species was targeted, not necessarily harvested.

[b]Comparable data on the net economic value of wildlife viewing trips made by nonresidents became available July 29, 1997, which was too late for inclusion in this report.

that price; this is known as the closed-ended or discrete-response format. Both formats were used in the course of the ADFG survey.

This is what is known in economics as a contingent valuation question. Contingent valuation generally is the process of measuring monetary value by eliciting it directly through some form of survey. It is an alternative to valuation through inference from behavior, such as the travel cost model approach mentioned above. Its use for this purpose was first suggested by Ciriacy-Wantrup (1947), who suggested that, just as people vote on taxes or bonds for state and local government programs, they could be asked how they would vote or what they would be willing to pay for something promoting the use or protection of the environment. This approach was first implemented by Davis (1964) to value recreation in the Maine woods. With the use of travel cost, it has become the main method of environmental valuation (Carson and others 1990; Carson and

Mitchell 1989). In recent years, there has been some controversy about the reliability of contingent valuation to measure what are called nonuse values—these are discussed further below (Diamond and Hausman 1994 and Hanemann 1994). But most critics view contingent valuation as valid for measuring the value of activities like recreational behavior (Desvousges and others 1993).

The estimates of WTP in table 6.1 are broken down by wildlife species, by type of recreation (hunting, primary nonconsumptive, and secondary nonconsumptive), and by type of participant (resident and nonresident). For moose-hunting trips, the average consumer's surplus was $181 per trip for resident hunters and $393 for nonresident hunters—these are what hunters would have been willing to pay beyond what they actually paid. For residents viewing moose on a nonconsumptive recreation trip, the consumer's surplus was $123 per trip. For caribou hunting, the average consumer's surplus was $168 per trip for resident hunters and $432 for nonresident hunters; for residents viewing caribou, the consumer's surplus was $135 per trip. (It is not possible to tell from the information made available to the committee whether any of the differences in value between moose and caribou are statistically significant.) For wolf hunting, the average consumer's surplus was $1,500 per trip for resident hunters and $351 for nonresident hunters; for residents viewing wolves, the consumer's surplus was $195 per trip. These estimates of consumer's surplus per recreation trip are broadly consistent with other published estimates of the consumer's surplus from similar types of recreation. They are also broadly consistent with other types of information on the value of hunting in Alaska, such as what people are willing to pay for permits to hunt on private land; on some privately owned lands, use fees exceed $2,000 per hunter (Boyce and others 1995).

It should be emphasized that the values in table 6.1 are not differentiated by where the trip occurred or the type of hunting or viewing conditions (including animal abundance) at the site of the trip; they are averaged over different sites and different conditions of animal abundance. Thus, they should be thought of as estimates of the average consumer's surplus per trip for a typical trip of each kind. That has 2 important consequences. First, one cannot directly infer an estimate of the total consumer's surplus from wildlife viewing or hunting in Alaska. The consumer's surplus per trip almost certainly varies across trips: the last trip is generally less valuable than the first trip. If one multiplied the values in table 6.1 by the total number of trips of each type, this would almost certainly understate the total consumer's surplus from wildlife viewing or hunting in Alaska. Second, there would be some inaccuracy in using these values alone to estimate the recreational benefits of a changed pattern of wildlife viewing and hunting resulting from game management or predator control programs in Alaska. That is because the values in table 6.1 do not account for changes in the consumer's surplus per trip resulting from improved conditions at a given site and site substitution whereby a person wanting to hunt or view wildlife switches from a site that he would have visited before to another site that now becomes more attractive as a result of the increase in wildlife population at that location. When there is site substitution of this form, the gain is the increase in consumer's surplus resulting from having to travel less far because one takes advantage of a newly improved site or resulting from obtaining enhanced enjoyment at the site where the activity is pursued. This increase could be more or less than the existing average consumer's surplus per trip and cannot be inferred from it.

In addition to uncertainty about the increase in consumer's surplus per trip, there would also be some uncertainty about the impact of a game management or predator control program on the pattern of recreational behavior, both the total volume of wildlife viewing or hunting trips and the allocation of this total among alternative sites—uncertainty that can best be resolved with the aid of a model like the travel cost model described above.

Most existing analyses of game management or predator control programs appear to have used other approaches that are questionable. One unreliable approach is to assume that there is a net increase in recreation activity that is directly proportional to the increase in animal population—if the animal population rises by 10%, there will be 10% more viewing trips or hunting trips. That approach rests on a chain of assumptions, each of which is dubious:

• The assumption that if the animal population at a site increases, the attractiveness of that site for recreation increases in the same proportion—it becomes 10% easier for viewers to spot an animal or for hunters to catch one. Instead, however, there might be non-linearities or threshold effects that lead to a nonproportional impact on the attractiveness or quality of the recreation experience.

• The assumption that if the attractiveness of a site increases, visits to that site increase in the same proportion. However, consider 2 sites of equal attractiveness that split a recreation market 50:50. Suppose that one of the sites becomes perceptibly superior to the other because of a 10% increase in the chance of viewing wildlife there. Conceivably, the entire recreation market could now switch its patronage to the improved site, leading to a doubling in visits there. Moreover, the enhanced attractiveness of the recreation opportunity would conceivably induce new people to take up the activity or existing participants to increase their participation. Those are merely examples that illustrate the general point that the impact on visitation can be disproportionally larger or disproportionally smaller than the change in site attractiveness.

• The assumption that a proportional increase in visits to a site implies an increase in total participation in the activity. Some of the increase might represent substitution of existing trips away from other sites, as in the above example.

For all of these reasons, therefore, one cannot generally assume that the total volume of recreation activity will change in proportion to the total size of the animal population. For consumptive uses, such as hunting, a variant of that approach is to estimate an allowable increase in the total harvest on the basis of biological considerations and then assume a proportional increase in hunter effort (recreation activity). Alternatively, there could be retrospective data on harvest by hunters, but not on hunter effort. Boertje and others (1996) used retrospective data on the increase in harvests in areas with wolf control programs to infer an increase in hunting activity. But, that assumes that the catch rate (or its inverse, the amount of hunter effort per moose harvested) remains constant. A firm foundation for the assumption does not exist.

The approach of inferring the level of hunting activity from data on harvests is sometimes defended on the basis of the observation that the number of applications for special hunting permits greatly exceeds the available supply. It is argued from that that the availability of animals is the limiting factor on hunting activity and that therefore the level of hunting activity should increase in proportion to the size of the allowed harvest. It is certainly true that when access to particular types of hunting is regulated through some sort of quota system, such as a special permit, it is very likely that there will be some increase in hunting activity if the allowed harvest is increased. But there is still the question of substitution: some of the increased hunting activity might represent a diversion of trips away from other types of hunting or outdoor recreation. Moreover, the magnitude of the excess demand for regulated permits could yield a distorted impression of the true size of the "reservoir" of frustrated latent demand for hunting. If applying for a permit and not getting one is relatively inexpensive, and if a hunter is at all risk-aversive, he will apply for more permits than he actually expects to use. For example, he might apply for 4 permits but expect to take only 1 or 2 trips.

In this connection, some of the people who testified to the committee linked the decline in the sale of resident hunting licenses in Alaska over the last decade to the decline in moose and caribou populations in some areas favored by hunters. There might be such a causal connection, but it is likely to be confounded by some other factors—over the last decade, there has been some decline in participation in hunting throughout the United States, including areas where animal populations have not declined. Some of the decline reflects a generational change in habits and interests away from hunting and other consumptive uses of wildlife. And some of it might reflect increasing demands on people's time that make hunting more difficult for them. The presence of such confounding factors makes it all the harder to predict how much of an increase in hunting to expect if animal populations are increased.

Even when there are direct retrospective data on changes in the volume of hunting activity, there can be important pitfalls of interpretation. Suppose, for example, that there are 2 areas, 1 in which a wolf control program is conducted and the other with no wolf control program. Suppose, further, that there are data on moose-hunting activity in the 2 areas that show that hunting declined in the area with no wolf control, but remained roughly constant in the area with wolf control. Reid and Janz (1995) confront a scenario like this on Vancouver Island.

They compute the benefits of wolf control for black-tailed deer hunting by assuming that without the program the volume of deer hunting in the wolf-control area would have declined at the same rate as it did in the no-wolf-control area. They calculate the extra days of deer-hunting activity in the wolf control area by using this assumption and multiply the result by an estimate of the consumer's surplus per day of deer hunting. Their analysis requires an assumption that the 2 areas are sufficiently similar that any difference in deer hunting can be attributed solely to the presence of the wolf control program in 1 area. More important, the analysis ignores the question of substitution among hunting sites. It is possible, for example, that some or all of the increase in moose hunting in the wolf-control area is simply a diversion of hunting activity that otherwise would have taken place in the no-wolf-control area. In that case, although there is still a gain in utility for deer hunters, it is likely to be some fraction of the consumer's surplus per hunting trip rather than the full consumer's surplus, as assumed by Reid and Janz. What should have been calculated is the increase in consumer surplus per deer-hunting trip diverted from other areas to the wolf-control zone.

In summary, the 2 core problems in assessing the economic benefits of any increase in ungulate populations for hunting and wildlife viewing are to predict the change in recreation behavior—both the total level of recreational activity and its allocation among sites—and to estimate the increase in consumer's surplus from recreation that is associated with this change in activity. Neither of those can be assessed with biological models of animal populations. Instead, both require a model of people's preferences and behavior along the lines of the travel cost model mentioned above.

It might be useful to give an example of how such a model can be used, on the basis of application of the ADFG model of sportfishing in south-central Alaska for a case study of the economic impact of closing the Kenai River to sportfishing for king salmon during the last week of July (Carson and others 1990). Fishing for king salmon on the Kenai River accounts for about 8.5% of all fishing at that time of year by residents of south-central Alaska. It was estimated, however, that the closure would reduce their number of fishing trips by only about 1.5%, because most of them would still go fishing but switch to other species and other sites. With the closure, it was predicted that salmon-fishing trips would fall from about 53.5% of all resident fishing trips at that time of year to about 50%. Among salmon-fishing trips, anglers would switch from king salmon to other types of salmon. Anglers still choosing king salmon as their target species would switch to other sites; for example, the share of king salmon trips going to the Kasilof River on the Kenai Peninsula would increase from 11% to 45%, and the share of king salmon trips to the ocean off Deep Creek on the Kenai Peninsula would increase from 4.5% to 13%. The main impact of the change is a reallocation of trips to other species and sites, and the loss of consumer's surplus is smaller than it would be if the main impact were a reduction in total fishing itself. If the impact were a reduction in total fishing, resident anglers' loss of consumer's surplus from closing the Kenai River to king salmon fishing in the last week of July would amount to about $1.2 million; with the substitution of other species and sites, the loss drops to about $482,000.

Gain from Increased Recreation of Nonresidents Involving Prey Species

Several thousand nonresidents visit Alaska each year to hunt—at least 7,162 in 1991. At the conceptual level, the economic issues that arise in estimating a monetary measure of the gain in utility for nonresidents who benefit from improved recreation for prey species, such as hunting and wildlife viewing, are the same as those discussed above for residents. The core problems are to predict the change in recreation behavior—both the total amount of their recreation in Alaska and its allocation among alternative sites—and to estimate the change in their total consumer's surplus from recreation in Alaska. The actual change in behavior, however, and the resulting change in consumer's surplus are likely to be quite different for nonresidents and residents because they are different types of people and engage in different types of recreation. For a nonresident, visiting Alaska to hunt or to view wildlife is a fairly substantial undertaking that might occur only once or twice in a lifetime. At home, most of a hunter's trips might be 1-day or 2-day trips; but he will plan a much longer visit to Alaska, probably involving hunting at several locations. Moreover, most or all of the hunting trips at home are without a guide, whereas nonresidents commonly hire a guide to hunt in Alaska; indeed, it is required by state law for nonresident hunting of brown bear, Dall's sheep, and mountain goats. The transportation costs and the use of a guide make hunting in Alaska much more expensive for nonresidents than for residents. On a brown bear hunting trip to Kodiak Island, for example, ADFG found that nonresident US citizens spend an average of $9,545 per trip, of which nearly $6,500 is for the guided hunt, about $1,000 for transportation, and another $1,000 for taxidermy (Loomis and Thomas 1992). A study by Yu (1991) found that 90% of the clients served by guides in Alaska were nonresidents and 92% of guides' revenue came from nonresidents.

Because hunting and wildlife viewing in Alaska are different types of recreational experiences for nonresidents and residents, it is necessary to estimate separate models of recreation participation and site choice for nonresidents and residents and to conduct separate analyses of the recreational impacts of a predator control program. ADFG has a separate model of sportfishing by nonresidents in south-central Alaska (Carson and others 1990) but not for other recreational activities.

Loss from Reduced Recreation Involving Predators

The effect of a predator control program on resident or nonresident recreation involving predators needs to be factored into analyses with the recreational impact involving prey species. The general issues are to predict the change in the level of recreation participation, the impact on site choice, and the resulting change in consumer's surplus. In the case of predators, however, some types of recreation might be affected differently—viewing of wolves might be reduced whereas, depending on how the control program is implemented, hunting of wolves increases. The reduction in opportunities to view wolves could be important, in that ecotourism and wildlife viewing are the fastest growing sectors of the Alaska travel industry. For example, more than 40% of tourists to Alaska cite wildlife viewing as important to them; 10% report being frustrated by not seeing the numbers and types of wildlife that they had hoped to see. At the Alaska Wolf Summit, Bob Dittrick, of Wilderness Birding Adventures, stated: "A large percentage of the inquiries we receive ask about the potential of seeing wolves. And many place it as their main personal goal for the trip even though I am selling birding trips. The tourists coming to Alaska want to see all the different wildlife, but from my experience wolves top the list. Given the choice, these people would give up seeing lots of wildlife for the chance of seeing one wolf. This is also true of Alaskans who vacation in the state." As noted earlier, data from the Boundary Waters Canoe Area and Glacier and Yellowstone national parks suggest that the presence of wolves, bears, and other large carnivores is a major attraction for visitors. Moreover, the sport-hunting value of brown bears, largely frustrated in the lower 48 states by Endangered Species Act restrictions, could motivate a small number of trophy hunters to visit Alaska if additional opportunities were provided.

The information available to the committee suggests that, in planning the wolf control programs in 1993 and 1995, ADFG did not conduct any analysis of the economic impacts on wolf-related recreation by residents or nonresidents.

Nonuse Values for Non-Native Residents and Nonresidents

In addition to the impact on both recreation involving both predator and prey species and on personal income from employment and profit in Alaska, to be discussed later in this chapter, people might be affected by predator control programs in various other ways that reflect their concerns and involvement with these animals. Use of the animals as a source of food or hides will be discussed in the following section, but both residents and nonresidents might experience a gain or loss of utility from a predator control program for reasons separate from their interest in recreation, meat, hides, or other such uses of the animals. For example, they might care about the prey species and want it to be preserved, regardless of whether they themselves plan to view it, hunt it, or eat it. Or they might care about the well-being of Native peoples whose way of life depends on the prey species. If people feel like that, they will experience a gain or loss of utility from a predator control program that must be accounted for in an economic analysis, just as gains or losses associated with impacts on recreation, food, and personal income are counted.

In economics, this is known as existence value, nonuse value, or passive-use value—the value that people place on an item for motives that are separate from their interest in using it. The concept was first suggested by Krutilla (1967), who observed that people with no interest in using an item might still believe that it should be preserved and be willing to pay something to ensure its preservation. Their value will not be reflected in their use of the item, but it could be reflected in other forms of behavior, such as engaging in political or charitable activities or engaging in a consumer boycott to express displeasure. Consequently, existence value cannot be discerned by analyzing people's demand to use the item through the revealed preference approach described above. Conventional market data are inadequate to measure this value. Instead, one must use an approach that can directly elicit people's preferences.

In the case of predator control in Alaska, evidence strongly suggests the existence of substantial nonuse values, at least for some segments of the resident and nonresident populations. The widespread public support for wolf protection and restoration in the lower 48 states and a general perception of wolves and brown bears as imperiled species, as well as the public controversy about wolf control within Alaska, point to that. However, the magnitude of the nonuse values which exists for both prey and predator species, and includes concern by non-Natives for the well-being of Native peoples in the absence of predator control, is unknown.

Some critics have argued that nonuse values cannot be reliably measured through contingent valuation surveys. The question was considered by a panel of experts, headed by Nobel laureates Kenneth Arrow and Robert Solow, convened by the National Oceanic and Atmospheric Administration (NOAA). The panel asked to advise NOAA on whether contingent valuation is capable of providing reliable information about nonuse values in the context of assessing damages from oil spills. The panel concluded that contingent valuation studies can produce estimates reliable enough to be the starting point for a judicial or administrative determination of natural-resource damages, including lost nonuse value (Arrow and others 1993). It stipulated that, to be acceptable for this purpose, contingent valuation studies should adhere to guidelines dealing with various aspects of survey design and implementation.

No contingent valuation study on nonuse values associated with predator control has been conducted in Alaska. However, a contingent valuation study on nonuse values associated with wolf recovery in Yellowstone National Park was commissioned by the US Fish and Wildlife Service for its environmental impact statement on wolf reintroduction into Yellowstone and central Idaho (Duffield and Neher 1996). A telephone survey was conducted of 2 random samples of households, a national sample and a sample drawn from the 3-state region (Idaho, Montana, and Wyoming) where wolf reintroduction was to occur. Because of the divided nature of opinion on the matter, an initial screening question was used to elicit whether the respondent favored or opposed efforts to reintroduce wolves. Depending on the response, two separate lines of questions were asked. Those in favor of wolf reintroduction were asked whether they would be willing to buy membership in a trust fund being established to support wolf reintroduction; those opposed were asked whether they would be willing to buy membership in a trust fund being established to oppose wolf reintroduction. In each case, a discrete-response format was used—respondents were told that membership in the trust would cost a specific amount and asked whether they would be willing to pay it. From the yes or no responses, the researchers derived an estimate of the mean or median WTP to promote or oppose wolf reintroduction among members of the population.

The survey found considerable differences of opinion nationally and in the local region. In the local region, opinion was quite closely divided—49% favored wolf reintroduction, 43% were opposed to it, and 8% did not know. National opinion was more lopsided—57% favored wolf reintroduction, 29% were opposed to it, and 14% did not know. The estimate of mean WTP was higher among supporters than opponents of wolf reintroduction and higher in the local region than nationally. Estimated mean WTP for wolf reintroduction among supporters was $20.50 per person locally and $8.92 nationally; estimated mean WTP to oppose wolf reintroduction among opponents was $10.08 per person locally and $1.52 nationally. As a further precaution, the researchers reduced those values by about 70% to allow for survey evidence that, with charitable contributions, people typically pay only about 30% of what they say they would pay.

Two other findings from the study of wolf reintroduction at Yellowstone are that use values are a minor and unimportant component of the benefits; instead, the conclusion is determined by nonuse values. Second, the geographic scope of the analysis and whose values are taken into account make a substantial difference. If the analysis is restricted to the local region, the costs of wolf reintroduction in fact outweigh the benefits, and the program would not be economically justified. If the analysis is broadened to the national population, the costs stay the same but the benefits are greatly expanded, and the conclusion overwhelmingly supports wolf reintroduction.

The conclusions of a study for Alaska are unlikely to be the same as those for Yellowstone, because the circumstances are different. In particular, there is no counterpart at Yellowstone of the impact on Natives' cultural values and lifestyles, which could turn out to be an important component of the evaluation of predator control in Alaska. As with Yellowstone, though, the conclusions of a study in Alaska are likely to be sensitive to the geographic scope of the analysis and the determination of whose values should be taken into account.

Impact on Non-Native Residents' Personal Income from Employment and Profit

According to a survey conducted in December 1992 just before the wolf control program was suspended, around 80% of Alaskans thought that tourism and the visitor industry are critical to the well-being of the state (AWRTA 1995). Also, 40% of residents believe that the negative effect of the boycott on the state's tourism industry was a sufficient reason to cancel the program. The national boycott of tourism in Alaska certainly appears to have played a role in Governor Hickel's decision to suspend wolf control in 1992. The actual effect of the boycott on the Alaskan economy might have been quite small, but there were 2 mitigating factors: the governor moved quickly to stop the wolf control program, and it was ended before the height of the tourist booking period, which is mid-January through February. Surveys of tourist-related businesses conducted during and after the boycott suggest that, had it continued for longer, it would have caused a substantial reduction in tourism in 1993.

Depending on the circumstances and the method of implementation, a similar boycott of tourism in Alaska is likely if a wolf control program is again authorized. Because the situation is so variable, it is impossible to predict with precision the extent of a future boycott or the magnitude of its effect on the state's economy. Nevertheless, even if a boycott turned out to have only a modest effect on the economy, one cannot know this in advance. From an economic point of view, the measure of loss is the net reduction in personal income, whether from employment or from profit. Some of the tourist expenditures represent costs of produced inputs rather than personal income from employment or profit. In the case of sportfishing in south-central Alaska, which was estimated to generate expenditures in the state by residents and nonresidents amounting to about $93 million annually, only about $18 million was personal income in Alaska (Carson and others 1990). In addition to that direct impact, there are indirect and induced impacts in the rest of the Alaska economy, estimated for sportfishing to amount to another $47 million of personal income in Alaska. Those estimates are subject to the comments above about marginal versus average coefficients and offsetting changes in economic activity elsewhere in the economy. The same would apply to recreation losses from a tourist boycott. To the extent that jobs and income lost from a tourism boycott are made up by jobs and income gained in the activities that tourists substitute for a trip to Alaska, the net loss is reduced. If the substitution occurs out of state, from an Alaskan perspective there is a loss; from a national perspective, there is not.

Costs of Predator Control Program

In addition to any loss of personal income from changes in tourism and any loss of utility for residents or nonresidents associated with reduction in predator populations, there are costs to government agencies and others associated with planning and implementing a predator control program. ADFG has prepared summaries of its costs for previous wolf control programs, but not specifically for the 1992 or 1994 program.

Three points should be noted about the calculation of such costs. First, the relevant concept is the increase in cost due to the program, that is, its marginal cost. Second, in this context opportunity costs might be more important than direct costs. As noted earlier, for an agency with a generally fixed budget and staff, part of the increased cost of providing a new activity might be the other services that have to be reduced or postponed when agency personnel are diverted to work on the new activity. To the extent that a program creates troublesome precedents for an agency or limits its future range of options, these too are opportunity costs associated with the program. Third, if fees are used to finance a program, they must be treated with caution because fees might be a transfer, rather than a real social cost.

ECONOMIC IMPACTS OF PREDATOR CONTROL ON NATIVE AND SUBSISTENCE PEOPLES

This section examines the possible impacts of wolf control on rural Alaskans, those living in generally isolated communities where subsistence activities can be an important element of the community's economic base. About 20% of Alaska's population—about 124,000 people in 1995—lives in rural areas (Wolfe 1996). Those people reside in about 225 communities, most of which are off the road network and have fewer than 500 people. Only about half the rural population is Alaska Native, but most small villages are predominantly Alaska Native.

This section begins by considering how rural Alaskans might be affected by predator control. It requires an overview of the social, economic, and cultural setting for this population—particularly the importance of subsistence activities. The next part examines the economic value of wildlife resources to rural Alaskans. The section ends with a discussion of those issues from other (noneconomic) social-science perspectives, in particular applied anthropology.

How Rural Residents Are Affected

Rural Alaskans support themselves through fishing, hunting, gathering of wild foods, and small-scale cash employment. Wolfe (1984, 1996) has characterized this as a mixed subsistence-cash economy. Wildlife resources are more than food sources and part of the economic base of these communities. It should be noted that from the Alaska Native perspective, subsistence is not just taking food from the local environment, but is an essential part of the Alaska Native way of living and entails a wide range of practices and beliefs, including spirituality, community, sharing, connection to and respect for other life, connection to the seasons, and teaching the young and moving them into their full inheritance

(Merculieff and Soboleff 1994). Predator control affects rural residents directly and in several ways; not only are prey species, such as moose and caribou, used and valued, but so are the predators themselves. The impacts on urban residents can be usefully categorized as impacts on consumptive or nonconsumptive recreational uses and passive use, but the impacts of wildlife management on some rural residents are more pervasive, in that wildlife-related subsistence activities constitute their life and livelihood.

Residents of rural Alaska make greater use of wild foods than communities anywhere else in the United States. Wolfe (1996) estimated that about 43.7 million pounds of wild food are harvested each year in rural Alaska. That amounts to 375 pounds (usable weight) per person annually in rural Alaska and 22 pounds per person in Alaska's urban areas (Wolfe and Bosworth 1994). By comparison, the average American uses about 222 pounds of store-bought meat, fish, and poultry each year. Wolfe and Walker (1987) summarized the data from the ADFG Division of Subsistence harvest surveys in 98 of the rural communities. Per capita harvests vary widely from relatively urbanized communities like Kenai (38 pounds) to remote villages like Hughes on the Koyukuk River (1,498 pounds). The most productive areas are along the interior rivers and the Arctic Coast. The composition of harvests varies widely across the state, from predominantly caribou and sea mammals in the north to salmon and moose in interior villages.

Fishing and hunting for subsistence provide an economic base for many rural regions. Case studies of rural Alaska's mixed subsistence-cash economy include Behnke (1982), Fall and others (1986), and Wolfe (1981,1984). Similar analyses have been applied to the Canadian North (Usher 1981). In those economies, fishing and hunting are the central activities in the community and are conducted by family groups with efficient small-scale technologies, including fishwheels, gill nets, motorized skiffs, and snowmobiles (Wolfe and Walker 1987). Subsistence is augmented by cash employment commonly in commercial fishing, trapping, public-sector employment, services, and construction. The cash income enables families to purchase snow machines, fishing nets, rifles, and ammunition. The monetary incomes typically earned in villages are not large enough to support the family unless a portion is used in subsistence fishing and hunting (Wolfe 1991).

Subsistence resources are clearly important as food and an economic base in rural Alaska, but they also have a cultural meaning. One aspect of this is that these economies differ markedly from most modern market economies in terms of distribution. Harvests of subsistence resources are not sold in markets but are distributed on a kinship, status, role, or other basis to people in the community and nearby communities. Sharing is an important basis for reciprocity relationships, status, and identity.

Subsistence is broadly integrated through links with traditional beliefs and practices. Each village has a unique "seasonal round" of activity that is driven by the natural cycle of availability and quality of food. In the interior, this can include the greening of the first plants, return of waterfowl, early king salmon runs, summer runs of chum and sockeye, berrying, and fall runs of silver salmon. In the fall, moose are fat and in prime; caribou might be migrating through. The winter can be a time when pelts are prime and trapping is most productive. Subsistence practices as a system combine many aspects of life, such as work and sport, that are separate for most contemporary Americans. Hensel (1994) reports that when older Yup'ik men in western Alaska were asked "What do you do for fun?" and "What do you do for work?" one generally got the same answer: "hunting and fishing."

Some villages are more dependent than others on species that can be affected by wolf or bear predation. Predation by wolves or bears has minimal impact on major food sources for coastal communities that depend largely on marine resources or river-based communities with abundant runs of anadromous fish. But some communities are potentially vulnerable to the influence of predation by wolves or bears. Stokes and Andrews (1982) provide a discussion of the importance of moose to Nikolai and Telida, small villages in the upper Kuskokwim with populations of about 90 and 30 people, respectively, in 1981. At that time, Telida had no store, and the store in Nikolai depended on delivery by air from McGrath. In the fall of 1981, 52 moose were harvested by the 37 households of these 2 communities. As in most years, moose harvested in the 1981 season made up the major portion of the winter protein needs of people living in the area. Moose is, by far, the largest and most consistent and culturally important source of protein in the local food supply (Stickney 1981).

Residents expend a substantial cost in terms of both effort and money to acquire moose, in some cases up to 20% of annual household income just for operating expenses (primarily gas for boats). As Stokes and Andrews note, the prominence of moose at traditional ceremonial occasions, such as potlatches, is an indication of the special cultural values ascribed to moose. Caribou populations in the early 1980s were quite low in the area. Salmon contribute to the local diet but are in relatively poor condition in the upper Kuskokwim as they approach the spawning areas. In addition, ADFG regulations in the middle 1960s eliminated the traditionally used king salmon fence and trap, and that has resulted in lower harvests (Stokes 1983). Stickney (1980:11) summarized the importance of moose to Nikolai as follows: "Because moose is the most important item in the village diet the villagers will tolerate an expenditure of their limited cash resources to subsidize their hunting venture. This is not a luxury to them, but rather a necessity they don't question. Without a moose they must fall back on the store's expensive commodities without the sufficient cash base to provide a sufficient protein . . . diet. Due to supply problems the store cannot always be relied on to provide a nutritious adequate diet even if the cash were available."

Economic Values of Subsistence Use

As Klein (1989) has observed, fish and wildlife are essential to the subsistence way of life of many northern peoples. It is obvious that without these local food sources, the subsistence communities could not exist. The "importance" of these resources includes their economic contribution to the communities that use them and the more subjective psychological well-being "derived from a sense of economic security, and the cultural traditions or spiritual values that are interwoven with the resources" (Klein 1989: 99).

In principle, the economic value of an incremental change in these resources is the equivalent in monetary income that would make a given person just as well off. The problem is to find a method to measure the equivalence. For marketed resources, the value of the last unit of a good that is consumed can be closely approximated by the market price. A person is observed to have made the tradeoff of money and the good, and one can infer that the good is worth at least as much as the price of the last unit taken. However, many or most subsistence resources are not traded in established markets. Subsistence resources are distributed by sharing, rather than sale, so local markets do not exist. Aside from problems of isolation, there are laws in most states, including Alaska, that prohibit the sale of most wildlife resources.

Even where some important resources are available in markets (such as markets for salmon or halibut), the resulting prices are generally not relevant for those who do not participate in these markets. This includes most residents of rural Alaskan villages. The price of king salmon in Anchorage provides little information on what the salmon is worth to a resident of Nikolai and is perhaps no more relevant than the price of sockeye salmon in Tokyo. Nonetheless, the most common approach to valuing subsistence resources is to use the replacement cost based on market prices. That is clearly inappropriate, in that the cost of the replacement might have little or nothing to do with the actual change in economic well-being that a person realizes by having or not having the given resource. Although the approach is in principle wrong, estimates based on it have been used in the context of environmental impact assessment (Usher 1976), in litigation (Duffield 1997), and descriptively (Wolfe and Bosworth 1994). For example, Wolfe and Bosworth (1994) assume a replacement cost of $3-$5 per pound and estimate the replacement cost of all wild food harvests in Alaska at $131.1-$218.6 million. Although those numbers are questionable, they indicate the relative importance of wild foods to Alaska Natives. The per capita cash value of subsistence foods in the rural interior is estimated to be $3,063 per person, or about half the 1990 per capita income for Native families of $6,205. Of course, for any analysis of a change in management, the relevant issue would be not total values, but the marginal value of changes.

The problem with interpreting replacement cost as the marginal value of a subsistence resource is compounded by the problem of choosing the appropriate replacement or substitute. Many subsistence products have no commercially available equivalents, and available "substitutes" might be judged inferior on the basis of taste, texture, amount of fat, and possible presence of pollutants (Hensel 1994). The use of a range of $3-$5 per pound in some applications is based on retail prices of beef or salmon, which might or might not be equivalent to the resource at issue. In the *Exxon Valdez* case, the issue of using exact replacement of all subsistence resources, including marine mammals, led to average prices in 1994 of $10-$14 per pound in the settlement of the Alaska Native claim (Duffield 1997).

Brown and Burch (1992) reviewed the sparse literature on the economic value of subsistence harvest. They suggested a microeconomic model of subsistence harvests that has 2 components of value: the demand for participation in the activity and the demand for the product. Whether or not that is a useful way to frame the issue, the problem of estimating the components remains. The authors reviewed the range of possible methods, including travel cost and contingent valuation, and concluded that the latter, at least in principle, could provide relevant measures.

Wolfe and Walker (1987) analyzed a database that included both per capita subsistence harvests and income. They developed a model that shows a systematic tradeoff between those 2 components of the mixed cash-subsistence economies and provided an interpretation concerning factors that affect economic development in Alaskan rural villages. Not surprisingly, in areas where subsistence harvests are quite high, cash incomes tend to be low and vice versa. That is consistent with what one would expect because people have a fixed amount of time to allocate to different activities and the types and value of available work differ among communities. Duffield (1997) used the Wolfe-Walker data to calculate the marginal value of subsistence harvests based on the observed tradeoff of income and subsistence harvest. That approach, which used wage differentials across communities as a method for valuing site specific amenities, yielded estimated values of subsistence harvests in Alaska of about $30 per pound.

These analyses assume that Native peoples who live by subsistence hunting and fishing do so by choice, rather than out of necessity, and that they perceive themselves better off than if they were fully absorbed into the market economy. There is some support in the literature for that perspective on the part of Alaska and northern Canada Natives (Berger 1985; Klein 1989; Usher 1976). Hensel (1994: 10) provides anecdotal evidence for this view in the following quotation from an interview with a Yup'ik woman professional in Bethel in July 1992:

"You know, I have had some really good offers, more money and advancement if I would move to Anchorage or Juneau, but I've wanted to stay here so that I could continue to do subsistence.

"One time when my supervisor was out from Anchorage, we were having her over to dinner and serving her moose, strips, salmonberry and blueberry pie. I told her how important it was for me to be able to eat this food, and hunt and fish. I thanked her for all her help with my career but told her that I wanted to stay here."

Other Social Science Perspectives

As the preceding discussion suggests, converting changes in individual welfare to monetary measures for subsistence users is difficult. Assistance can be provided by other disciplines, such as cultural anthropology, that take account of the profound cultural differences associated with wildlife and wildlife management. In fact, anthropological studies are a valuable approach used by ADFG Division of Subsistence for evaluating management issues.

The committee has attempted to identify the marginal value of additional information on the basis of the likelihood that it would change a management decision. Several examples of ADFG Division of Subsistence community-level studies by that criterion have proved valuable in leading to apparently positive management changes. This type of social and economic analysis might be most appropriate for the remote rural communities because of the relatively homogeneous and distinct cultures in these communities; the relative autonomy and self-governance of these communities due to isolation, history, and the legal definition of subsistence rights to wildlife resources; and the recognition and acceptance by the larger society of the unique cultural and economic setting of the communities.

The ADFG Division of Subsistence, which was established around 1980, is unique in ADFG in that social science studies are important in its work. Some of these studies use an economic perspective (for example, Wolfe and Walker 1987) but a number of community-specific harvest surveys, which are very different from the Division of Wildlife Conservation hunter surveys, have been conducted. Hunter surveys focus on the primary game animals (moose, caribou, and so on) and provide a time series for harvests and hunter participation for all licensed hunters from 1983 to the present (Robert J. Walker pers. comm. 1997). In contrast, the Division of Subsistence studies are for *all* subsistence resources (from roe on kelp, to gull eggs, to berries, and so on), are at the community level, and typically collect data for 1 year only. Thus, the Division of Subsistence surveys provide a detailed picture of all subsistence activities for one time for the given community.

In addition to the harvest surveys and related reports, the division also has undertaken social-impact studies that focus on a specific resource in a given region, sometimes from the perspective of cultural anthropology. Reports of 2 of those studies are discussed here as examples of social science studies that have led to wildlife management changes in the communities of interest. The methods and findings might have elements of benefit-cost or other economic analyses but are predominantly cultural studies in which the behavior at issue appears to be largely motivated by ethical or cultural values. The first study summarized below concerns brown bear use in northwestern Alaska by village residents. It provides an example of where management changes were based almost entirely on cultural considerations and financial concerns were not an issue. The second study concerns changes in moose-hunting regulations in the upper Kuskokwim River drainage (part of GMU 19D), particularly affecting the villages of Nikolai and Telida. Here, the change involved resolution of conflict between subsistence users and fly-in hunters and was made mainly on equity grounds. Both of those policy changes appear to have been in the public interest, and the social science analysis, although it involved no formal economic analysis, was relevant and appropriate to the problem at hand.

Brown Bear Use in Northwestern Alaska

Loon and Georgette (1989) examined subsistence use of brown bears by residents of northwestern Alaska, particularly the Kotzebue Sound region (GMU 23) and to a lesser extent Norton Sound (GMU 22). The use of brown bears for food and raw material was prevalent in all the inland study communities, but coastal communities rarely used bears for food, because the flesh of coastal bears, which feed on carcasses of sea mammals has an unpleasant flavor. The authors note that northwestern Alaskans have an extensive array of traditional laws and lore regarding human-bear interactions. The traditional practices cover hunting strategies, butchering processes, personal conduct, methods of defense, and appropriate attitudes. Because brown bears are believed to have keen hearing, Inupiaq hunters do not openly discuss their bear hunts. There are believed to be severe consequences to the hunter and his family if these laws are not heeded. Hensel (1994:2), in a study for the Association of Village Council Presidents in western Alaska, makes similar observations for Yup'ik, noting that "the hunting of brown bears, and even discussing these animals, is potentially dangerous."

Loon and Georgette discuss the incongruity between the brown bear hunting regulations in place at the time and the customary and traditional hunting practices. The state regulations in the late 1980s presume that the primary use of a brown bear is as a trophy, whereas subsistence hunters' rules presume that the primary use is as a source of food and raw material. For example, state regulations in the late 1980s required that a person who kills a bear must personally present the skin and skull to an authorized representative of the ADFG for sealing within 30 days after taking. A person with a bear had to keep the skin and skull together until a representative of the department removed a rudimentary premolar tooth from the skull. In contrast with the treatment of most big-game species, discarded brown bear meat is not considered waste under the regulations.

Almost all of these brown bear hunting regulations were at odds with customary subsistence hunting practice. Some subsistence hunters leave the head in the field at the kill site as a sign of respect, a practice that is in conflict with the sealing requirements. Requiring hunters to salvage the skin and skull does not accommodate those who hunt for meat and fat only; to subsistence hunters *not* requiring the salvage of bear meat is objectionable. A bag limit of 1 bear every 4 years is inconsistent with the fact that in most villages only a small number of men actually harvest bears, and these hunters share their harvests throughout the community. The strong prohibition on speaking openly about brown bears includes not even speaking about one's intentions to hunt. Requiring local hunters to purchase a hunting tag before hunting brown bears, and thereby deliberately making their intentions known, is incompatible with traditional hunting practices. The regulatory hunting seasons at that time, April 15 - May 25 and September 1 - October 10, also conflicted with traditional hunting times, which can begin as early as March when some bears first come out of their dens. Those many differences between the hunting regulations and traditional practices are culturally based, "learned differences which derive from the social values of the hunter's community" (Loon and Georgette 1989:49).

Not surprisingly, given the mismatch between the regulations and traditional practices, Alaska Natives generally ignored the regulations. As a result, the state was able to collect only very limited information on harvest and use of bears. Loon and Georgette estimate that of the bears killed by interviewed hunters over the previous decade, only 3% were reported. They concluded that hunters would be more likely to report their bear harvests if regulations accommodated their hunting practices and if the reporting procedure were simple.

The study by Loon and Georgette led to changes in the management of bear hunts in some areas (Hensel 1994). For example, a western Alaska brown bear management area (covering essentially the Yukon-Kuskokwim delta and the Aniak and Togiak drainages) was established as a cooperative agreement of the Association of Village Council Presidents and the US Fish and Wildlife Service. In that area, tagging and sealing requirements were suspended after the fall of 1991. Residents were required to register in advance of brown bear hunting but paid no fee. The season was extended to run from September 1 to May 31.

Subsistence Resource Use in the Upper Kuskokwim

In the early 1980s, the ADFG Division of Subsistence produced a series of reports concerning subsistence use in the upper Kuskokwim River drainage, particularly communities of Nikolai and Telida (Andrews and Stokes 1984; Stickney 1980, 1981; Stokes 1984; Stokes and Andrews 1982), which depend heavily on subsistence resources. The studies began in 1980 when the McGrath local Fish and Game Advisory Committee asked the Board of Game to implement a controlled-use zone around the 2 villages. According to Stickney (1980), the proximity of the region to the municipalities of Anchorage and Fairbanks created a situation in which the game resources, especially moose, were subject to competition from urban hunters, as well as boat hunters from down-river

Kuskokwim communities. In Stickney's opinion, hunting by outsiders, annual village take, marginal moose habitat, and a high wolf population all acted to keep moose at a low density in the area. Moose was the most important food item in the local diet, and alternative subsistence resources (salmon, whitefish, and caribou) were not plentiful enough in the area to be dependable buffers for inadequate moose harvests. Stickney's assessment of the situation was that the survival of the villages was at stake. With regard to Nikolai, "the villagers hunt for meat for the winter and they will not return empty handed if possible even if their prey does not conform to the State's regulations." With regard to Telida, "in this case . . . the regulations imposed by the State have apparently little bearing on what the village of Telida faces as a reality. The villagers will ensure their own survival even if the regulations compel them to operate outside the permitted system." Stickney concluded that as of 1980 there were not enough data to make firm recommendations to resolve the problem.

At the spring 1981 Board of Game meeting, a controlled use area for moose harvest was established for part of GMU 19D, partly on the basis of the report presented by division staff (Stickney 1980, 1981). Stokes and Andrews (1982) monitored the controlled use area during the 1981 season and concluded that probably no aircraft entered the region for hunting. The institution of the controlled-use area was looked on favorably as a management tool by local residents. In 1981, the season was also changed to include a winter hunt in late November and early December. Stokes and Andrews noted that some traditional spring hunting of moose is also important for spring and summer food needs. In March 1982, the Board of Game adopted an extended winter moose season in portions of the controlled-use area from December 1 through February. Stokes (1983) and Andrews and Stokes (1984) evaluated the 1982-1984 winter moose seasons and concluded that they were successful in providing Nikolai residents the opportunity to harvest moose legally during times compatible with local environmental conditions and with cultural needs.

Sustainable Use of Wildlife by Alaskans and the Global Environment

Harvest of wildlife by Alaskans on a sustained yield basis for subsistence use, or as a replacement for store purchases of foods, presents an admirable model of human compatibility with the world environment through its minimal contribution to depletion of the world's nonrenewable energy and mineral resources, and the environmental pollution associated with their use (Harbo 1993). This is in contrast to most people in developed countries who are dependent upon food produced by intensive agriculture. The amount of energy and minerals consumed in the production of commercial food stuffs, especially in the United States, represents a major component of all nonrenewable resource use. This includes production of farm equipment, use of petroleum products for fertilizer production, operation of farm machinery, processing of foods, their storage, and shipment to markets. The contributions from agricultural production to pollution of the atmosphere (including "greenhouse" gases), the waters, and lands of the world are immense. Alaska's contribution to the nation's energy needs through the production of oil from the North Slope reserves is appreciated by many Americans. However, most Americans are not aware of the contribution that Alaskans make to the conservation of energy and other resources through their sustainable harvest and use of wild meats.

SOCIAL AND ECONOMIC IMPACTS IN RELATION TO DECISION-MAKING ON WOLF CONTROL

In its review of the social science methods that can be used to assess the social and economic impacts of a predator control program and by describing what is known about those impacts in relation to wolf control in Alaska, the committee identified major gaps in existing information. The next chapter discusses the general context in which wildlife management decisions must be made and the constraints on those decisions.

REFERENCES

ADFG (Alaska Department of Fish and Game). 1994-95. Trapper questionnaire statewide report.

ADFG (Alaska Department of Fish and Game). 1994. Subsistence in Alaska: 1994 Update. Division of Subsistence, Juneau, AK.

Alaska Department of Labor. 1996. Labor Department Estimates Population Trends. Research and Analysis Section.

Alaska Wildlife Alliance. 1992. Survey among Alaska residents regarding wolf hunting.

AWRTA (Alaska Wilderness Recreation and Tourism Association). 1995. Results of economic survey of AWRTA tourism business members on the effects of the tourism boycott. AWRTA, Wolf Issue/Tourism Boycott: Survey Results.

Anderson DB. 1995. The Alaska Department of Fish and Game public opinion survey on predator control in Game Management Unit 19D East. Report to the Alaska Board of Game. Division of Subsistence and Division of Wildlife Conservation, Alaska Department of Fish and Game, Juneau, AK.

Andrews ELand J Stokes. 1984. An overview of the Upper Kuskokwim controlled use area and the use of moose by area residents, 1981-1984. Technical Paper No. 99. Alaska Department of Fish and Game Division of Subsistence.

Arrow K, R Solow, E Leamer, P Portney, R Hadner, and H Schuman. 1993. Report of the NOAA panel on contingent valuation. Federal Register 58: 4601-4614.

Arthur LR, E Gum, E Carpenter, and W Shaw. 1977. Predator control: the public viewpoint. Pp. 135-155 *in* Transactions of the 42nd North America Wildlife Conference.

Bath AJ. 1987. Attitudes of various interest groups in Wyoming toward wolf reintroduction in Yellowstone National Park. MA thesis, University of WY, Laramie. 123 Pp.

Bath AJ. 1991. Public attitudes in Wyoming, Montana and Idaho toward wolf restoration in Yellowstone National Park. Pp. 91-95 *in* Transactions of the 56th North American Wildlife and Natural Resources Conference. Wildlife Management Institute, Washington, DC.

Berger TR. 1985. Village journey: the report on the Alaska Native Review Commission. Inuit Circumpolar Conference. New York: Hill and Wang.

Biggs JR. 1988. Reintroduction of the Mexican wolf into New Mexico—an attitude survey. MA thesis. NM State Univ, Las Cruces, NM. 66 Pp.

Boertje RD, P Valkenburg, and ME McNay. 1996. Increases in moose, caribou, and wolves following wolf control in Alaska. J Wildl Manage 60:474-489.

Boyce JR, DW McCollum, and JA Morrison. 1995. An economic impact analysis of the big game hunting guide industry in Alaska.

Braithwaite A and S McCool. 1988. Social-normative influences and backcountry visitor behavior in occupied grizzly bear habitat. Symp Soc Sci in Res Mgt, Univ Ill., Urbana, IL.

Brown TL, DJ Decker, and DL Hustin. 1981. Public attitudes toward black bears in the Catskills. New York State Department of Env. Conservation. Albany, NY. Pp. 108.

Brown TC and ES Burch, Jr. 1992. Estimating the economic value of subsistence harvest of wildlife in Alaska. *In* GL Peterson and others, Eds. Valuing wildlife resources in Alaska. Boulder: Westview.

Burghardt GM, RD Hietala, and MR Pelton. 1972. Knowledge and attitudes concerning black bears by users of the Great Smoky Mountains National Park. Int Conf Bear Res Mgt 2:255-273.

Buys C. 1975. Predator control and rancher's attitudes. Environ & Beh 7:81-89.

Carson RT and RC Mitchell. 1989. Using surveys to value public goods: the contingent valuation method. Johns Hopkins University Press, Baltimore, MD.

Carson RT, WM Hanemann, and D Steinberg. 1990. A discrete choice contingent valuation estimate of the value of Kenai King Salmon. J Behavior Econ 19:1-15.

Ciriacy-Wantrup SV. 1947. Major economic forces affecting agriculture with particular reference to California. Hilgardia; a journal of agricultural science 18:2-21.

Clawson M and J Knetsch. 1966. Economics of outdoor recreation. John Hopkins University Press, Baltimore, MD.

Colorado Division of Wildlife. 1989. Attitudes about wildlife and black bears. Denver, CO.

Davis RK. 1964. The value of big game hunting in a private forest. Trans N Am Wildl and Nat Resour Conf. 29:393-403.

Desvousges WH, AR Gable, RW Dunford, and SP Hudson. 1993. Contingent valuation: the wrong tool to measure passive-use losses? Choices: the magazine of food, farm and resource issues. Second Quarter 8: 9-11.

Diamond PA and JA Hausman. 1994. Contingent valuation: is some number better than no number? J Econ Persp 8:45-64.

Duffield JW and CJ Neher. 1996. Economics of wolf recovery in Yellowstone National Park. Trans. 61st N Am.Wildl and Nat Resour Conf 61:285-292.

Duffield JW. 1997. Nonmarket valuation and the courts: the case of the Exxon Valdez forthcoming Contemporary Economic Policy.

Dunlap T. 1988. Saving America's wildlife. Princeton: Princeton University Press. 222 Pp.

Fall JA, JC Schichnes, M Chythlook, and RJ Walker. 1986. Patterns of wild resource use. *In* Dillingham: Hunting and fishing in an Alaskan regional center. Technical Paper No. 135. Alaska Department of Fish and Game, Division of Subsistence.

Freeman AM. 1993. The measurement of environmental and resource values. Washington, DC: Resources for the Future.

Gardner C and K Taylor. 1992. Results of the wildlife management survey in the upper Tanana/ Fortymile region. Alaska Department of Fish and Game. Fairbanks, AK.

Hanemann WM. 1994. Valuing the environment through contingent valuation. J Econ Persp 8: 9-44.

Harbo S. 1993. Environmental sanity: think globally, act locally. Paper presented at the Wolf Summit. Fairbanks, Alaska, January 1993.

Hastings B. 1986. Wildlife-related perceptions of visitors in Cades Cove, Great Smoky Mountains National Park. PhD dissertation. Univ TN, Knoxville.

Hensel C. 1994. Brown bear harvests in the western Alaska brown bear management area, 1992/1993: Statistical Information and Cultural Significance. Manuscript.

Herrero S. 1970. Man and the grizzly bear. Biosci 20:1148-1153.

Herrero S. 1978. People and grizzly bears: the challenge of coexistence. Pp 167-179 *in* CM Kirkpartrick, Ed. Wildlife and people. The 1978 John S. Wright Forestry Conference. Purdue Res Foundation, West Lafayette, IN.

Hook R and W Robinson. 1982. Attitudes of Michigan citizens toward predators. *In* F Harrington and P Paquet, Eds. Wolves of the world. NJ: Noyes Publishing.

Hotelling H. 1931. The economics of exhaustible resources. J Polit Econ 29:137-75.

Jobes PC. 1991. The greater Yellowstone social system. Cons Biol 5:387-394.

Jonkel CJ. 1975. Opinion—of bears and people. Western Wildlands 2:30-37.

Jope K and B Shelby. 1984. Hiker behavior and the outcome of inter-actions with grizzly bears. Leis Sci 6:257-270.

Kellert SR. 1980. A national study of American attitudes, knowledge, and behaviors toward wildlife: Phases I, II, III. US Govt Printg Off, Supt Doc 024-010-00-625-1/2/4.

Kellert SR. 1985. Public perceptions of predators, particularly the wolf and coyote. Biol Conser 31:167-189.

Kellert SR. 1986. The public and the timber wolf in Minnesota. Transactions of the 51st North American Wildlife and Natural Resources Conference. Wildlife Management Institute, Washington, DC.

Kellert SR. 1991. Public views of wolf restoration in Michigan. Transactions of the 56th North American Wildlife and Natural Resources Conference. Wildlife Management Institute. Washington, DC.

Kellert SR. 1994. Public attitudes toward bears and their conservation. *In* JJ Clear, C Servheen, and LJ Lyons, Eds. Proceedings of the Ninth International Bear Conference. U.S. Forest Service, Missoula, MT.

Kellert SR. 1996. The value of life: biological diversity and human society. Island Press, Washington, DC.

Klein DR. 1989. Northern subsistence hunting economies. Pp.96-111 *in* RJ Hudson and others, Eds. Wildlife production systems. Cambridge University Press, New York.

Krutilla JV. 1967. Conservation reconsidered. Amer Econ Rev 4:777-786.

Llewellyn L. 1978. Who speaks for the timber wolf. Transactions of the 43rd North American Wildlife Conference. Wildlife Management Institute, Washington, DC.

Loomis JB and MH Thomas. 1992. Pricing and revenue capture: Converting willingness to pay into state and private revenue. Pp.255-274 *in* GL Peterson and others, Eds. Valuing wildlife resources in Alaska. Westview Press, Boulder, CO.

Loon H and S Georgette. 1989. Contemporary brown bear use in northwest Alaska. Alaska Department of Fish and Game, Division of Subsidence. Technical Paper No. 163.

Lopez B. 1978. Of wolves and men. New York: Scribner's.

Matthiessen P. 1988. Wildlife in America. New York: Viking.

McCollum DW and SM Miller. 1994. Alaska voters, Alaska hunters, and Alaska nonresident hunters: their wildlife related trip characteristics and economics. Alaska Department of Fish and Game.

McConnell KE, IE Strand, and NE Bockstael. 1990. Habit formation and the demand for recreation: issues and a case study. Pp. 217-235 *in* A Link, Ed. Advances in applied micro-economics, Volume 5. JAI Press.

McCool S, A Braithwaite, and D Cikanek. 1990. Beliefs about bears. Wildl Soc Bulletin.

McNamee T. 1984. The grizzly bear. New York: Alfred A. Knopf.

McNaught DA. 1987. Wolves in Yellowstone Park? Park visitors respond. Wildl Soc Bulletin 15:518-521.

Merculieff L and W Soboleff. 1994. Letter to the Alaska Federation of Natives and the participants of the Elders and Youth Conference from participants in the Alaska Native traditional knowledge and ways of knowing workshop, September 13-14, 1994.

Miller SM and DW McCollum. 1994a. Alaska voters, Alaska hunters, and Alaska nonresident hunters: their characteristics and attitudes towards wildlife. Alaska Department of Fish and Game.

Miller SM and DW McCollum. 1994b. Tradeoffs between uses and users in natural resource management: A case study of who gains, who loses, and what it means. Alaska Department of Fish and Game.

Miller SM, R Sinnot, and D McCollum. 1994c. Differentiating among wildlife-related attitudinal groups in Alaska. Transactions of the 59th North American Wildlife and Natural Resources Conference. Pp. 512-523.

Miller SM, SD Miller, and DW McCollum. 1997. Attitudes toward and relative values of Alaskan brown and black bears to resident voters, resident hunters, and nonresident hunters. Int Conf. on Bear Res and Management, Vol 10.

Murray A. 1975. Influence of education programs on wolf conservation in Canada. In DH Pimlott, Ed. Wolves. Supplementary. 43 Pp. World Conservation Union, Morges, Switzerland.

Naess A and I Mysterud. 1987. Philosophy of wolf policies. I: general principles and preliminary exploration of selected norms. Cons Biol 2:16.

Peek JM and others. 1991. Restoration of wolves in North America. The Wildlife Society, Technical Review 91-1. Wildlife Society, Washington, DC.

Pelton M, C Scott, and GM Burghardt. 1976. Attitudes and opinions of persons experiencing property damage and/or injury by black bears in the Great Smoky Mountains National Park. Int Conf Bears Res Mgt 3:157-167.

Petko-Seus PA and MR Pelton. 1984. Backpacker bear knowledge in Great Smoky Mountains National Park. Hum Dim Wildlife Newsletter 3:8-12.

Petko-Seus PA. 1985. Knowledge and attitudes of campers toward black bears in Great Smoky Mountains National Park. MS. thesis, Univ TN, Knoxville, TN.

Reading R, TW Clark, and SR Kellert. 1994. Attitudes and knowledge of people living in the greater Yellowstone ecosystem. Soc and Nat Resour 7:349-365.

Reid R and D Janz. 1995. Economic evaluation of Vancouver Island wolf control. Pp. 515-521 in LN Carbyn, SH Fritts, and DR Seip, eds. Ecology and conservation of wolves in a changing world. Canadian Circumpolar Institute, Occasional Publication No. 35, Edmonton.

Shepard P and B Sanders. 1985. The sacred paw: the bear in nature, myth, and literature. Viking Penguin, Toronto, Canada.

Stickney A. 1980. Subsistence resource utilization: Nikolai and Telida—interim report. Technical Paper No. 20. Alaska Department of Fish and Game, Division of Subsistence.

Stickney A. 1981. Subsistence resource utilization: Nikolai and Telida—interim report II. Technical Paper No. 21. Alaska Department of Fish and Game, Division of Subsistence.

Stokes J and E Andrews. 1982. Subsistence hunting of moose in the Upper Kuskokwim controlled use area, 1981. Technical Paper No. 22, Alaska Department of Fish and Game, Division of Subsistence.

Stokes J. 1983. Winter moose season in the Upper Kuskokwim controlled use area, 1982-1983. Technical Paper No. 72. Alaska Department of Fish and Game, Division of Subsistence.

Stokes J. 1984. Natural resource utilization of four Upper Kuskowin communities. Technical Paper No. 86. Alaska Department of Fish and Game, Division of Subsidence.

Stuby R, E Carpenter, and L Arthur. 1979. Public attitudes toward coyote control. USDA Economic Statistics and Cooperative Service. ESCE-54. Washington, DC.

Trahan RG. 1987. Wilderness user's attitudes, information use and behaviors in relation to grizzly bear dangers in the backcountry. Yellowstone National Park, National Park Service.

Trice AH and SE Wood. 1958. Measurement of recreation benefits. Land Econ 34:195-207.

Turner RK, D Pearce, and I Bateman. 1993. Environmental economics. Johns Hopkins University Press, Baltimore, MD.

Usher PJ. 1976. Evaluating country food in the northern native economy. Arctic 105-20.

Usher PJ. 1981. Sustenance or recreation? The future of native wildlife harvesting in northern Canada. Pp. 56-71 *in* Proceedings, First International Symposium on Renewable Resources and the Economy of the North, M Freeman, Ed. Assoc of Canadian Universities for Northern Studies, Ottawa.

Wolfe RJ. 1981. Norton Sound/Yukon delta sociocultural systems baseline analysis. Technical Paper No. 59. Alaska Department of Fish and Game, Division of Subsistence.

Wolfe RJ. 1984. Commercial fishing in the hunting-gathering economy of a Yukon River Yup'ik Society. Etides/Inuit Studies Supplemental Issue 8:159-183.

Wolfe RJ and RJ Walker. 1987. Subsistence economies in Alaska: productivity, geography, and development impacts. Arctic Anthropology 2:56-81.

Wolfe RJ. 1991. Trapping in Alaska communities with mixed subsistence-cash economies. Technical Paper No. 217. Alaska Department of Fish and Game, Division of Subsistence.

Wolfe RJ and RG Bosworth. 1994. Subsistence in Alaska: 1994 update. Alaska Department of Fish and Game, Division of Subsistence, 4 Pp.

Wolfe RJ. 1996. Subsistence food harvests in rural Alaska and food safety issues. Paper presented to the Institute of Medicine. National Academy of Sciences Committee on Environmental Justice, Spokane, WA.

Young S. 1946. The wolf in North American history. Caldwell, ID, Caxton Printers.

Yu X. 1991. Valuing the impact of Alaska's guiding industry on the state's economy. M.Sc. thesis, University of Alaska, Fairbanks.

7

Decision-Making

Science progresses by developing and testing hypotheses. Managers who wish to use the best scientific information when making management decisions often need to generate new data. Thus hypothesis-testing is also a component of natural resource management. However, decision-making, not hypothesis-testing is the most important task of a natural resource manager. A manager must make decisions about how to manipulate a variety of entities—including wildlife, habitats, and regulations that govern consumptive and nonconsumptive use of wildlife—on the basis of available information. Time, personnel, and funding are usually insufficient to obtain all the information desired by the manager when a decision must be made. And a decision delayed is a decision made.

Ideally, there is feedback between management and hypothesis-testing. Information should be updated and revised as the results of decisions are observed, a process known as *passive adaptive management. Active adaptive management* involves designing management decisions so that their implementation maximizes what can be learned to guide future management decisions. Ideally, resource management is combined with continuing experiments designed to generate new knowledge.

That characterization exaggerates the distinction between science and resource management because the design of scientific experiments involves active decision-making. Indeed, the statistical literature on hypothesis-testing and experimental design explicitly adopts a decision-making framework. Nevertheless, the distinction is important.

THE GENERAL DECISION-MAKING FRAMEWORK

A successful decision is one that achieves its goal efficiently and without causing undesirable side effects. Decision-making involves both selection of a goal and selection of a method to accomplish it. The major issues of decision-making are *who* decides, *what* is being decided, *how* the decision is made, and *what* the costs and benefits of alternative options are. Estimating costs and benefits is often difficult because one must consider the probability of compliance, ease of enforcement, management costs, ability to deal with surprises, and costs of dealing with opposition to policies.

All decisions—whether for resource management or hypothesis-testing—involve a common general structure. The 2 key elements of the structure are the *choice set*—the alternatives that are to be considered—and the *objective function*—the criteria by which the choice among alternatives is to be made. Usually, various types of costs and benefits are associated with the alternatives; these are identified and defined by the objective function. Decision-making identifies and chooses the alternative that gives the best value of the objective function, subject to whatever constraints exist in the system.

For decisions involving continuous variables, rather than variables that are in discrete categories, this corresponds to evaluating marginal benefits and marginal costs. In other words, the manager evaluates the costs and benefits of changes in policies. Examples include evaluating the consequences of extending or reducing a hunting season, increasing or decreasing bag limits, and increasing or decreasing the intensity of predator control or management in an area. In deciding how much effort to allocate to generating new data, a manager needs to assess the likely marginal value of new data, that is, the contribution of increases in the database to the foundation on which decisions are made.

To make wise decisions, natural resource managers (all decision-makers, for that matter) need to avoid two ways in which a poor decision is likely to be made. One way to make a poor decision is to fail to identify the best among the alternatives considered. That is what people usually refer to when they say that a "wrong" decision was made. Alternatively, a poor decision can be made because the decision-making process itself is flawed. A flawed decision-making process overlooks some relevant alternatives (that is, the best alternative is not among those being considered), uses an incorrect objective function (that is, poor criteria for judgment are used), or overlooks some relevant objectives or constraints (for example, it ignores political constraints or incorrectly characterizes the ecological traits of the target species).

Both pitfalls pose dangers for natural resource managers. Inadequate or bad data can lead to poor characterization of alternatives and failure to recognize important constraints. The need to make decisions quickly can lead managers to reduce the number of options considered, and so to exclude the best possible options. Legislative mandates or public pressures can force managers to exclude valuable options from the set they consider.

Today, major wildlife management decisions in Alaska are made under intense scrutiny by a public that holds a wide range of views about which management goals are appropriate and how they should be achieved. Consensus about goals and methods is unlikely to be reached soon, but decision-making can be carried out in a way that is perceived to be fair. Perception of fairness is the critical ingredient that leads people to accept what they consider to be unfavorable decisions and to be willing to advance their views and preferences within the decision-making framework rather than attacking it from the outside.

The following pages review the constraints under which wildlife management decisions are currently made in Alaska and evaluates processes that might assist in the making of wise decisions that have broad public support.

CONSTRAINTS ON WILDLIFE MANAGEMENT DECISION-MAKING IN ALASKA

Quantity and Quality of Data

Biologists and managers in the Alaska Department of Fish and Game (ADFG) have done a credible job of assembling and interpreting data on populations of wolves, bears, moose, and caribou in Alaska. Indeed, some of the long-term data sets on wildlife populations in Alaska are among the best available anywhere. ADFG scientists regularly publish their data in peer-reviewed literature, so their data and methods of analysis are carefully scrutinized by scientists elsewhere. In game management units where predator control actions are being considered, special efforts are made to gather more-extensive data to evaluate the desirability of predator control and its likely ecological consequences.

However, the database that supports management decisions is inadequate because resources, both human and financial, are insufficient to gather data intensively over an area the size of the state of Alaska, and, as pointed out previously, unpredictable environmental perturbations prevent accurate predictions of the outcome of any control effort even when extensive data sets are available. Therefore, the public must be willing to accept uncertainties in the outcomes of all management decisions, including predator control. To demand predictive certainty is biologically unrealistic.

Politically Imposed Constraints on Decision-making

The depth of conviction with which different views about goals and methods of wildlife management are held by Alaskans powerfully constrains management goals and methods of achieving them. Current constraints include prohibitions on same-day aerial hunting and legislation that mandates maximization of human harvests of moose and caribou as the primary objective of wildlife management. In combination, existing political constraints seriously limit both the alternatives that can be considered and the methods that can be used for achieving whatever options are chosen. Although the committee recognizes that the public, through its legislative bodies, appropriately sets overall policy goals, current political constraints on decision-making are so severe that the best options are unlikely to be among those that are considered. Good decision-making is highly unlikely under these conditions.

Constraints Imposed by Patterns of Land Ownership

Although less than 1% of land in Alaska is privately owned, the complex distributional pattern of federal, state, and tribal lands (figure 6.1) generates a complex set of problems. For the most part, political boundaries do not coincide with ecological boundaries. In addition, individuals of the target species regularly move across political boundaries. As a result, appropriate and effective wildlife management decisions seldom can be made within the boundaries of single political units. To make a wise decision, the geographic scale of a problem to be solved must be identified, and all relevant data and groups of people must be assembled.

The need to make decisions on relevant spatial and temporal scales is clearly recognized by ADFG, and the procedures generally used by ADFG and Board of Game are among the most thorough and open anywhere in the United States. The fact that the existing process, despite its generally favorable structure, has not been able to generate broadly supported decisions illustrates the vulnerability to disruption of processes in the current decision-making environment.

The most complex jurisdictional issue is posed by the highly fragmented distribution of Native lands in Alaska. These units are too small to constitute adequate areas within which to develop biologically appropriate management plans. But, indigenous knowledge of wildlife populations in those areas is extensive, and indigenous people have sought more control over wildlife management decisions within their lands. How indigenous knowledge and political desires should be incorporated into the development and implementation of management decisions is considered in the next section.

INCORPORATING INDIGENOUS KNOWLEDGE INTO DECISION-MAKING

People who spend time observing animals where they live accumulate rich stores of knowledge about them. Local knowledge is clearly useful for biologists who conduct surveys that cover vast areas. While people trained in Alaskan Native traditions might have different methods of data collection than those trained in Western scientific methods, it is not difficult to conceive of strategies for integrating these methods.

However, traditional knowledge also involves less tangible, but equally important issues—those involving social values—such as, what constitutes ethical treatment of animals. Cultural traditions sometimes clash. For Alaskans of European descent, removing the tooth from a dead bear's skull is largely a matter of indifference; for Alaskan Natives, it is an affront to an animal that they hold in great respect. And yet state regulations in the 1980s required hunters in Alaska to remove and deliver the tooth of any bear they killed to state game officials. In this case, when policymakers understood how important this issue was, the regulations were modified.

Ideally, policymakers will be informed of social issues and be able to avoid policies that clash with social values. Most bear hunting regulations in the 1980s were at odds with customary Native hunting practices. Hensel (1994) estimated that consequently only 3% of bear harvests were reported (see section in chapter 5 on "Brown Bear Use in Northwest Alaska" for further discussion). Clearly, policies that deviate from cultural norms are difficult to implement and often simply disregarded. This, in turn, erodes both the respect of people for and their willingness to participate in political institutions.

Attention to traditional knowledge has three important facets. First, there is the potential for better information about the current status of wildlife populations. Second, insight into traditional knowledge can promote the development of more effective wildlife management policy. Finally, equity and respect for other cultures requires attention to different traditions of knowledge.

A workshop on "Alaska Native Traditional Knowledge and Ways of Knowing," held in Anchorage on September 13-14, 1994, put forward the following list of issues to be included among the principles or policies governing the use of Native traditional knowledge:

- Research should be defined by the community.
- People should not be required to participate in research that violates their ethics, values, or spirituality.
- All participants in programs, projects, or initiatives should be treated as equals.
- Indigenous cultures should not be mandated to use outdated technology while harvesting wildlife only because it is traditional.
- Alaska Native cultures and ways of life should be the foundation for self-regulation.
- Alaska Natives should be active participants in management of fish and wildlife resources through co-management options, with adequate funding provided to allow such structures to function.

Alaska Natives believe that co-management is the primary method by which

Native knowledge should be more fully incorporated into natural resource management. The term co-management refers to the sharing of responsibility for management functions by indigenous peoples, government, and other local organizations. The balance of authority and responsibility can vary considerably. At one extreme would be government control over management decisions, with limited input from indigenous peoples. Indigenous control, with limited input from the government, marks the other end. Between the extremes, a rich array of possibilities for shared decision-making authority exists.

The 1994 amendments to the Marine Mammal Protection Act created a new section that authorized Alaska Natives and the federal government to enter into agreements that allow local organizations to manage the activities of hunters and to participate in research on marine mammals. Co-management under the act may range from simple agreements to discuss research topics to the development of a plan under which Alaska Native organizations conduct complete scientific investigations on their own or develop their own management plans with local enforcement.

In addition to its technical meaning, co-management has an important symbolic meaning, corresponding to the strongly held desire of Native tribes and organizations to exert control over the development and implementation of management schemes on their lands. Thus, co-management has a normative meaning that describes the level of authority and control that a community believes to be appropriate.

The most appropriate system of co-management in any specific situation depends on the types of "management" functions that are involved. These functions fall into 4 interrelated categories: research, regulation, allocation, and enforcement. Research includes the gathering of baseline biological and other environmental data, performing experiments, and gathering harvest data. Regulation involves restrictions on harvest seasons and locations, bag limits, methods of harvest, and which species or age and sex classes can be harvested. Allocation determines who is allowed to harvest the wildlife. Enforcement ensures that the applicable regulations are followed and identifies who is authorized to do the enforcing.

On May 8-9, 1995, the Rural Alaska Community Action Program (Rural CAP) hosted a discussion on "Co-management: Establishing Principles, Policies, and Protocols" in Anchorage. Participants in the workshop recognized that, depending on the species and systems being managed and the goals of management, co-management schemes could be addressed in a variety of ways. For example, individual communities could have co-management agreements for each species that they harvest. Alaska Native regions could have agreements covering all their communities and the species they harvest. Species-specific agreements could be developed that cover the entire range of a species, or special agreements could be made for populations of a species that are judged to be threatened.

Finally, agreements that are consistent with ecological boundaries might be appropriate.

Workshop participants agreed that co-management plans governing Native lands should be implemented by representatives of Alaska Native organizations designated by the tribes. Once selected, the representatives would meet to coordinate a comprehensive approach. The envisioned process would be a statewide one, in which ideas would come from local hunters and users, be evaluated through regional processes, and finally coordinated on the spatial scales judged to be most appropriate for the specific management goals.

The Alaska Native community believes that effective conservation of fish and wildlife is possible only through co-management agreements governed by the following principles (prepared on May 8-9, 1995):

• Alaska Native tribes should authorize the entities that enter into negotiations of co-management agreements.

• The federal government should abide by the tribes' choice and negotiate only with those Alaska Native tribes or tribally-authorized Alaska Native organizations.

• Co-management negotiations and agreements should be based on the principle that Alaska Native tribes, the federal government, and, for species under the jurisdiction of the state, the state government operate on a co-equal and government-to-government basis.

• Co-management agreements should recognize and incorporate Native traditional knowledge, ecosystem-based approaches, and stewardship of resources.

• Research conducted or funded by the federal or state governments on Native lands should follow criteria for research developed by the Alaska Native community.

• There should be periodic meetings between Native and high-level federal and state government officials to discuss and review co-management.

Co-management of natural resources requires people who might hold contrasting world views to devise management plans jointly. Conflicts can arise over how data about status and trends in animal populations can and should be gathered. For example, darting and anesthetizing animals and taking tissue samples, which are acceptable to Western biologists, might be regarded as being overly invasive by many Alaska Natives. Similarly, methods of monitoring harvests and obtaining data from harvested animals can be contentious.

MULTI-JURISDICTIONAL MANAGEMENT OF WOLVES, BEARS, AND THEIR MAJOR PREY

It is beyond both the mandate, and the expertise of the members of the committee to analyze in detail and make recommendations about the most appropriate management schemes for wolves, bears, and their ungulate prey in Alaska in cases where more than 1 political or administrative entity shares responsibility for determining and implementing management plans. Nonetheless, the ecological characteristics and requirements of the focal species suggest that local management regulations must be compatible with those of neighboring regions across which animals move. In some extreme situations, federal laws can pre-empt local management. However, because none of the species of concern is threatened or endangered, pre-emption of authority under the Endangered Species Act should not, in general, be necessary.

REFERENCE

Hensel C. 1994. Brown bear harvests in the western Alaska brown bear management area, 1992/1993: Statistical Information and Cultural Significance. Manuscript.

8

Conclusions and Recommendations

INTRODUCTION

In this chapter, the committee summarizes the results of its analysis of the biological, socioeconomic, and decision-making underpinnings of management of wolves, bears, and their principal mammalian prey (with focus on moose and caribou) in Alaska. The results are presented as conclusions; recommendations flow from many but not all of them. The committee makes no recommendations about whether predator control should be carried out. To attempt to do so would go well beyond its mandate. Whether and when to control or not control predators is a policy decision to be made by the Alaska Board of Game on the basis of input from the public and recommendations and data provided by state and federal agency personnel. The role of the committee is to advise on how scientific, socioeconomic, and decision-making data can best be used to assist managers to make wise decisions.

BIOLOGICAL CONCLUSIONS AND RECOMMENDATIONS

With respect to the biological bases of wolf and bear management, the committee was asked to evaluate 3 questions:

- To what extent do existing research and management data provide a sound scientific basis for wolf control and grizzly bear reductions in Alaska?
- To what extent does our current level of knowledge allow accurate pre-

diction of the effect of a predator control program on predator and prey populations?

• What critical data gaps exist in our scientific understanding about these populations, and what would be needed to fill them?

The first 2 questions pertain to the cumulative results of past efforts to gather data and perform experiments. The 3rd question asks about the directions of future research. Because the first 2 questions are really different ways of looking at the same data, we combine them into a single composite question:

• In attempts to understand interactions between moose and caribou and their habitats and predators, have appropriate types of data been gathered, and has enough been learned from past research to identify the information needed to enable us to predict quantitative responses of prey populations to predator control efforts?

The knowledge needed to make wise predator management decisions includes information on interactions between predators and their prey, on interactions between prey and their habitats, and on long-term population dynamics of the key species. The committee's biological conclusions are divided into 2 categories: broad conclusions that pertain to assessing population trends, and narrow conclusions that pertain to biological aspects of wolf control.

Conclusion 1: Wolves and bears in combination can limit prey populations.

The committee concludes that there is clear evidence that under some conditions wolves and bears can keep moose and caribou populations suppressed for many years and that under appropriate conditions predator control can accelerate the recovery of prey populations. Alone or with bears, wolves can keep moose and caribou populations at levels below the carrying capacities of their environments. Reducing predator numbers can release this regulation temporarily.

Nonetheless, even in the cases of predator removal with the most extensive data, the evidence is insufficient to establish the existence of dual stable states, that is, one relatively stable state with high densities of both predators and prey and another with stable state with low densities of both. Moreover, field data are rarely sufficient for rigorous identification of how long prey densities are likely to remain high after predator control stops.

Conclusion 2: Wolf control has resulted in prey increases only when wolves were greatly reduced over a large area for at least 4 years.

There are only three cases, two of which have been conducted in Canada, in which sufficient data were collected to determine whether wolf and/or bear reductions caused an increase in adult populations of moose or caribou. Although there are confounding variables in each of the three cases, air-assisted wolf reduction for at least 4 years in fairly large areas apparently resulted in increases in caribou and moose calf:cow ratios followed by increases in adult populations. There have been no cases in which reduction of bear populations has resulted in increases in adult populations of moose or caribou. In many cases the experimental design was not adequate, the area was apparently too small, and/or the control effort not extensive enough. Our review of past attempts at wolf control indicates that it is likely to be successful when air-assisted wolf reduction is used over an area of at least 10,000 km^2; wolves are the primary predator of all age classes of the targeted ungulates; wolves are reduced to at least 55% of their precontrol numbers for at least 4 years; and the weather is favorable for ungulate survival. Under these conditions, moose and caribou may increase—at least during the years of control and perhaps longer. If the above conditions are not met, reducing the number of wolves is unlikely to increase ungulate populations.

Predator management might have worked more often than this in the past, but there is no scientific justification for the thesis that it has. This conclusion is not a criticism of the ADFG, which has often not been given the resources to use proper experimental designs for management actions. The sizes of areas treated and duration of treatments are often determined independently of the factors that influence wolf-prey dynamics. That many of the examples of unsuccessful predator control in Alaska were not well-designed and were a poor use of time and money as far as contributing to scientific understanding of predator-prey relations is to be expected in a program in which management, not research, has been the primary objective. We commend the ADFG for having been able to carry out basic research while simultaneously being responsive to requests from a variety of stakeholders.

Recommendation: Wolves and bears should be managed using an "adaptive management" approach in which management actions are planned so that it is possible to assess their outcome. That way managers can learn from the experience and avoid actions with uninterpretable outcomes or low probability of achieving their stated goals. Management agencies should be given the resources to conduct their management projects as basic research.

Conclusion 3: Expectations that managed populations in Alaska will remain stable are not justified.

People naturally prefer stable resources, and many consider this a reasonable goal for wildlife management. However, in northern ecosystems, such as those in Alaska, major population fluctuations are typical; stable populations are not. Natural fluctuations are the background against which management must work. Populations of prey are generally either increasing or decreasing when predator management begins. It is extremely difficult to tell how close a prey population is to the carrying capacity of its environment at a particular time. By perturbing the natural system, management could increase prey populations above the carrying capacity of the environment, and cause a deeper crash than would otherwise have taken place. That is apparently what happened in the 1970s after the massive predator control efforts of the 1950s.

Recommendation: Management objectives aimed at achieving stable populations of wolves, bears, and their prey should recognize that fluctuations in populations can be expected and provisions made for them. Before any predator management efforts are undertaken, the status of the predator and prey populations should be evaluated (including whether they are increasing or decreasing), and the carrying capacity of the prey's environment should be evaluated.

Conclusion 4: Data on habitat quality are inadequate.

Habitat quality is an important determinant of the dynamics of populations of large mammalian herbivores and omnivores. Predator control efforts are likely to succeed in increasing prey populations only if sufficient habitat of adequate quality exists to support the expanded populations—that is, only when prey populations are well below the environmental carrying capacity.

The primary deficiencies in current scientific understanding of what factors regulate and limit moose and caribou populations in interior Alaska and what management methods might be best are: (a) there has been too much emphasis on predator/prey relationships at the expense of other environmental factors. Research and experiments on diet, habitat relationships, and fire ecology are badly needed; (b) the ecology of both brown and black bears and their impact on moose and caribou are poorly understood; and (c) air-assisted wolf reduction and poisoning are the only methods that have been successful in the past, both are no longer socially acceptable, and little research has been conducted on alternative methods.

ADFG scientists have developed sound methods for monitoring wolves, moose, and caribou in the last 20 years, and more recently for bears. They have overemphasized the use of calf:cow ratios as an indication of ungulate population growth and should put more effort into measuring changes in adult population sizes. Nevertheless, they are to be commended for their many high-quality publications in peer-reviewed scientific journals. More complete data are needed, particularly about habitat quality for moose and caribou and their related physiological status, and about the ecology of bears. Short-term data dominate the existing database, but are insufficient for predicting long-term prey population trends.

Intensive studies of caribou-habitat and moose-habitat relationships, like those carried out on the Kenai, need not be duplicated in each area where there are plans to intensify management of caribou or moose. Nevertheless, a biologically justifiable scheme for intensive management involving reduction of predators of moose or caribou in a specific area must be based on data that suggest that the habitat can support an increased population. Current information is insufficient to allow reasonably quantitative predictions of probable increases in moose and caribou populations, because habitat quality estimates are rough and knowledge of bear population densities is poor. Attempts to increase moose or caribou populations in the past have been unsuccessful when knowledge of those factors was inadequate.

Recommendation: Research on alternative control methods should be adaptive, in the sense that it should be informed by the history of past successful and unsuccessful efforts. There should be more attention to experimental design and monitoring of results. Changes in both the population growth rate of the prey species and hunter satisfaction should be monitored.

Recommendation: ADFG should broaden the scope of its studies of predator and prey species. It should collect better data on habitat quality and on bear ecology. They should continue to increase its development of long-term data sets. Additional data are especially needed on bear foraging and population ecology, on quantitative and qualitative changes in habitats, and on the long-term consequences of predator control. The use of controlled fire should be further investigated as a tool for increasing the carrying capacity of moose habitat. Future research on these topics needs to be coordinated among the agencies that share jurisdictional authority over wildlife and wildlife habitats.

Recommendation: Collaborative relationships among ADFG and the land management agencies and jurisdictions should be strengthened so that habitat studies and habitat management efforts are well-coordinated.

Conclusion 5: Modeling of population dynamics will enhance the use of data already collected and enable more efficient use of limited resources.
Data collection on population trends and status of wolves, bears, caribou, and moose is expensive in terms of personnel, time, and equipment. Modeling of population dynamics offers a powerful means of analyzing and interpreting these data. With long-term data sets, modeling can greatly improve attempts to identify causal factors in predator and prey population trends. In the last decade, advances in computer technology have enabled the development of new methods for analyzing complex data sets, which have led to a revolution in the discipline of population modeling. Given the expense of data collection, modeling is a cost-effective way to use limited resources, and investment in expertise in population and resource modeling enhances the ability to synthesize and interpret field data.

Conclusion 6: Wolves, bears, and their prey are vulnerable to human actions but in different ways.
Although many people perceive wolves and bears as imperiled species, given the extent of natural habitats, the size and inaccessibility of much of Alaska,

unintentional extirpation of any of these species from large areas is extremely unlikely. There might be concern for wolves in some areas of southeastern Alaska, but these are not areas where wolf control has been contemplated. Although wolves were extirpated from the lower 48 states through habitat destruction associated with the expansion of agriculture, and through intensive poisoning, trapping, and hunting to protect domestic animals, natural ecosystems remain largely intact in Alaska, because agriculture and other land developments have been minimal, and species on which wolves can prey are abundant. The capacity of wolves for rapid recovery after population reductions remains unimpeded in Alaska.

Nevertheless, these predators and their prey differ dramatically in their vulnerability to human activities. Bear populations can be reduced quickly by liberalized hunting regulations where access is adequate; because of their low reproductive rates, they recover slowly. Furthermore, because bears are difficult to census it is particularly difficult to assess when populations are overharvested, another reason to manage them very conservatively. Even though bears kill many moose and caribou calves, it is not clear that reducing bear populations will cause an increase in the adult populations of moose or caribou.

Moose and caribou also have low reproductive rates and can easily be (and have been) overharvested. Sex-specific harvests, which result in highly skewed adult sex ratios, can also lead to reduced reproductive rates. But wolves have high reproductive potential and disperse widely such that their populations often can withstand annual harvest rates as high as 35% and keep their numbers stable from year to year. Where wolves have been suppressed, their populations have recovered rapidly. Bears, in contrast, have recovered slowly from suppression.

Recommendation: Wildlife policy makers in Alaska should be more sensitive to signs of overharvest and more conservative in setting hunting regulations and designing control efforts, particularly with respect to moose, caribou, and bears.

Conclusion 7: The design of most past experiments and the data collected do not allow firm conclusions about whether wolf and bear reductions caused an increase in prey populations that lasted long after predator control ceased.

Of the 11 cases of predator control analyzed by the committee, only 1 actually measured changes in hunter behavior, and only 2 provided evidence that prey densities increased in later years. (The committee did not review the many cases of predator reduction that lasted only a year.) Some of the others showed increases in fall calf:cow ratios, but those ratios might not be strongly correlated with later population increases. In the 2 experiments in which prey densities were known to increase, wolves were reduced by air-assisted shooting by more than 40% per year for more than 4 years over large areas, and the reductions were accompanied by both reduced hunting pressure on prey and mild winters. In the first case, GMU 20A, subsequent increases in moose and caribou lasted 16 and 14 years, respectively. However, because there was a simultaneous reduction in human harvest of moose and caribou, the conclusion that wolf control was responsible for the increase is uncertain. In the second case, in Finlayson, caribou densities increased after wolf control, but not enough time has elapsed to judge whether the increase will be more than temporary. Again, the experiment is confounded because human harvest of the prey species was greatly reduced at the same time. In addition there was a significant increase in mineral exploration after the wolf reduction ended. In other words, even in the 2 successful predator control efforts, the relative contributions of predator control and reduced harvests cannot be determined. The results of the other control experiments were either negative or so seriously confounded with other possible causes as to be uninterpretable.

Recommendation: Future experiments should be based on more thorough assessment of baseline conditions and should be designed so the causes of subsequent population changes can be determined.

Conclusion 8: Perfect prediction is unattainable.
Even if predator control experiments are well designed and based on extensive scientific data, consistent prediction of their results is, in principle, unattainable. That is because all estimates of population data are associated with errors and unpredictable events; droughts, severe winters, and fire can dramatically and quickly change environmental conditions. Therefore, both managers and the general public must accept that occasional failure to achieve intended goals will always be part of wildlife management actions.

Conclusion 9: Many past predator control and management activities have been insufficiently monitored.
Potentially valuable information that could have permitted more thorough evaluations and interpretations of results have often not been gathered. Many cases of predator control have suffered from incomplete treatments because of political interference. All control and management activities are expensive, so it is important to make full use of each one.

Recommendation: All control activities should be viewed as experiments with clear predictions. Control activities should be designed to include clearly specified monitoring protocols of sufficient duration to enable determination of whether the predictions are borne out and why.

SOCIOECONOMIC CONCLUSIONS AND RECOMMENDATIONS

With respect to the economic bases of wolf and bear management, the committee was asked to evaluate 3 questions:

• What existing economic studies and economic research methods can be used to evaluate the full economic costs and benefits of a predator control program for consumptive and nonconsumptive uses of wildlife resources in Alaska?

• What additional economic methods or additional data would be necessary for a comprehensive assessment of the economic costs and benefits of a predator control program?

• What strengths and limitations should be considered in assessing those economic analyses?

Conclusion 10: Benefit-cost analyses of management changes require at least three categories of information: biological relationships among predators, prey, and their environment; human behavioral response to changes in perceived quality of the use in question (for example, hunting success); and frameworks for valuing the change in use (or availability) of the resource.

Those data are needed because the goal of management programs is not simply to increase or decrease populations of animal species, but rather to increase the quantity and quality of human interactions with these species. Purely biological measures do not capture those values. The economic implications of management changes that affect wildlife (biological) populations can be properly understood only when all 3 values (biological, human, and monetary) are linked. Critical elements of these methods include defining the types of use at issue, the relevant user populations, and the spatial extent of the market for the resource's services. For example, use might not be limited to hunting by Alaska urban and/ or rural populations, but might include nonconsumptive (viewing, existence value) uses by populations both inside and outside Alaska. How the problem is correctly framed makes a substantial difference to the outcome of the analysis.

Recommendation: A set of studies should be commissioned to analyze hunting and viewing bear, wolves, moose, and caribou using techniques such as travel cost. This study should distinguish between values of Alaskan residents and nonresidents and Alaskan Natives and non-Natives. These studies should be designed both to provide information needed to determine the costs and benefits of different management policies and to provide a template for subsequent benefit-cost studies to be conducted by ADFG.

Conclusion 11: Evaluations of Alaska predator control programs have not gathered, analyzed, and assessed the full economic costs and benefits.

Existing socioeconomic studies have not measured benefits and costs for all relevant human populations, have used too limited or incorrect parameters for the user populations that have been examined, and have been too narrowly defined spatially to identify net effects in the spatial market of interest (for example, substitution and human response to management changes have not been measured). Analyses of the response by hunters and others to changes in wildlife populations have been quite limited. Densities of hunted animals are only one of several factors that influence hunter behavior and hunter success. Therefore, using only biological data on the treated population and a control population is insufficient.

Recommendation: ADFG should increase its efforts to evaluate human responses to management actions on spatial and temporal scales large enough to match the scale of the affected market. Travel cost models and contingent valuation should be applied to past and future management actions to improve assessments of value.

Conclusion 12: Social science research in Alaska is needed to support the design and evaluation of predator control experiments.

The Alaska Native cultures in rural Alaska have distinctive, diverse, and extensive mores and traditional practices concerning subsistence hunting and gathering. In several cases, applied anthropological research by the Division of Subsistence has led to changes in wildlife management (for example, bear management in northwestern Alaska). In some cases, particularly where cultural or ethical issues are dominant, the most appropriate social science perspective is not an economic one. The committee believes that collection and analysis of relevant economic, social, and cultural data could improve decision making by ADFG.

A social and economic impact assessment that generates information on the costs and benefits of particular decisions; analyzes the perceptions of various demographic groups regarding these decisions; and evaluates the cultural consequences of the decision will increase the likelihood that decisions are based on valid and reliable data and will probably prevent particular interest groups from unduly influencing decisions. This requires continuing data collection on human populations analogous to those obtained on wildlife populations. These data are equally as important as biological information in achieving wildlife management and policy-making that is effective and feasible. To be of optimal value, socioeconomic and cultural information needs to be obtained in a consistent, systematic, and comprehensive manner on all relevant stakeholders, and be routinely incorporated into the policy making process.

Recommendation: A formal procedure should be created, with adequate resources and trained personnel, to gather relevant economic, social, and cultural data and to incorporate this information into management and decision-making.

The specific tools of benefit-cost analysis and applied anthropology should be used in the analyses performed on those data.

Conclusion 13: Wildlife is, by definition, a public resource.

Under the constitution and statutes of the state of Alaska, the commissioner of fish and game is charged with "conserving and managing Alaska's fish and wildlife resources in the interest of the economy and general well-being of the state." In a democracy, all major societal interest groups should have the opportunity for meaningful involvement in decisions that affect them. In addition to the social impact assessment process recommended above, that requires continuing and formally organized opportunities for agencies to interact with various stakeholders. The perceived lack of substantive participation and involvement in management decisions by various citizen groups has been a major contributor to the current wolf and bear controversy. An extensive public involvement process might be initially be time-consuming and expensive, but in the long-run it is likely to generate more efficient and more acceptable decisions.

Recommendation: Procedures should be developed to allow the public to be substantively involved at all stages of the policy and regulatory decisions.

Conclusion 14: Greater potential for agreement may exist among Alaska's diverse constituency than is generally assumed.

The committee believes that a broad consensus exists among the Alaska public about the positive value of healthy populations of wolves, bears, and their prey. A broad consensus also exists about the appropriateness of predator control under conditions where prey are being kept well below the environmental carrying capacity and the scarcity of prey is adversely affecting subsistence users of ungulates. In the midst of substantial differences over the appropriateness of various means of predator control and how serious an "emergency" must be, there appears to be a basis for socially acceptable and sustainable policies.

Such policies could be developed if wildlife agencies and the public were involved in a broad strategic planning process that endeavored to achieve consensus regarding fundamental management goals. In the absence of strategic planning, division rather than cooperation, and crisis rather than aggressive management, tend to prevail. In the absence of long-term agreed upon goals it is also hard to allocate limited resources efficiently and equitably. An overarching articulation of agency priorities and directions would involve a long-term vision that is clearly communicated to the public and that is responsive to public concerns. Crisis management invites interference with public decision making and causes long term damage in many ways, such as erosion of public confidence, restriction of management options through uninformed directives, and loss of willingness of groups to work toward common solutions.

Recommendation: ADFG and the Alaskan public should engage in the development of a long-term strategic plan for the state's wildlife resource that is periodically revised as necessary.

Conclusion 15: Conflicts over management and control of predators are likely to continue indefinitely.

Even given a broad consensus on goals and improved public involvement in decision-making, differences will inevitably arise among competing constituencies regarding the allocation and management of wildlife resources. Differences in values and criteria for control of wolves and management of bears and ungulates might lead, in particular circumstances, to major conflicts. Therefore, formal procedures for dealing with conflicts will always be needed. ADFG does not have a conflict resolution strategy, but it has experience with using a facilitated conflict resolution process in the development of wolf control plans, and this experience could be drawn upon to establish a standard agency process for resolving conflicts that involve resource allocation and decision-making.

Recommendation: A formal conflict resolution process should be developed and adopted to help avoid the kind of intractable and wasteful dispute that has characterized the recent history of wolf and bear management in Alaska.

Conclusion 16: Decentralization of decision-making authority is not a panacea for solving wildlife management problems, but it is likely to be helpful in many circumstances, particularly in rural communities.

Many wildlife management decisions in Alaska are made at the level of GMUs or subdivisions of them. Such decisions must be based on detailed local information that supplements more general biological and social impact data. Therefore, effective and efficient decision-making needs to be customized by using local and traditional knowledge and targeted to the needs and interests of local constituencies. The great diversity of human populations in Alaska, their varied uses and perceptions of wildlife, and the tremendous variability of the Alaskan environment require that sort of management tailoring.

Nevertheless, decision-making cannot be completely decentralized because wildlife populations are affected by regional and national factors as well as local ones. Wildlife might migrate across jurisdictional boundaries and users often travel long distances to interact with wildlife. Those long distance moves of users and used affect many biological, sociological, and economic values underlying management policies. Moreover, people living far from sites of potential management actions might feel that they have a stake in those actions even if they do not expect to visit the areas in question.

Recommendation: Decision-making should be partly decentralized through formal consultation procedures whereby the views of local groups are solicited before decisions are made. In management situations involving rural and indigenous groups, more refined co-management decision-making structures should be developed where appropriate.

Conclusion 17: Interagency cooperation could improve management, reduce public confusion, and eliminate unnecessary duplication.

Wolves, bears, and related ungulate populations inevitably cross the jurisdictional boundaries of Native, state, and federal agencies. Because much of Alaska's land is under federal control, habitat and wildlife must be managed in cooperation with and with the approval of the federal government.

Recommendation: ADFG should assume a leadership role in strengthening cooperative agreements between the various jurisdictions and agencies involved in wolf and bear management in Alaska.

APPENDIX
A

Letter from Governor Tony Knowles Requesting Study

TONY KNOWLES
GOVERNOR

P O Box 110001
Juneau. Alaska 99811-0001
(907) 465-3500
Fax (907) 465-3532

STATE OF ALASKA
OFFICE OF THE GOVERNOR
JUNEAU

July 25, 1995

Dr. Bruce Alberts, President
National Academy of Sciences
2101 Constitution Avenue NW
Washington, DC 20418

Dear Dr. Alberts:

I am writing to request that the Academy consider undertaking a scientific review and economic analysis to help resolve a highly controversial and complex public policy issue in Alaska - the management of wolves and bears. The issues surrounding control and management of predators to increase the harvest of moose and caribou are of great interest and importance not only to Alaskans, but to people across the country and in many other nations.

Over the years, Alaska's game regulators have adopted numerous approaches to predator management. Wolf control measures have been implemented, and later dropped due to public outcry. The Alaska Legislature recently passed measures mandating that Alaska's Board of Game, the state's game regulating entity, provide for "intensive management" of game populations, including predator control. The Board and the Alaska Department of Fish and Game are currently trying to interpret that statutory directive and determine how to responsibly implement game management measures under this law. Another measure currently pending before the Legislature goes even further by authorizing a bounty on wolves.

When I took office late last year, I suspended the state's current wolf control program after it was revealed that the program was unacceptable in its treatment of both wolves and nontargeted species. As Governor, I believe I have a responsibility to see that Alaska's game management program meets three tests: it must be based on solid science, a full cost-benefit analysis must show that the program makes economic sense for Alaskans, and it must have broad public support. I will not reinstitute predator control measures unless and until they meet those three tests.

There always will be disagreement about predator control from ethical and other perspectives. However, public confidence in the science and socioeconomics upon which

Mr. Bruce Alberts
July 25, 1995
Page 2

management is based can go a long way toward public acceptance of a management program. Therefore, I request that the National Academy of Sciences consider overseeing a thorough and unbiased review of the biological and ecological science and an economic cost-benefit analysis of predator control in Alaska.

State officials would work with the Academy to develop the scope of the research and the specific questions which would provide the framework for these scientific and economic assessments. I welcome your expression of interest, as well as a preliminary timeline and cost estimate for each of the two studies.

As you have questions or require further information, please contact me directly, or Dr. Wayne Regelin, director of the Division of Wildlife Conservation at (907) 465-4190.

Sincerely,

Tony Knowles
Governor

Biographical Information on Committee Members

Gordon H. Orians, (*Chair*) is professor emeritus of zoology at the University of Washington, Seattle. He is an ecologist and environmental scientist who conducts research on the evolution of vertebrate social systems, the structure of ecological communities, plant-herbivore interactions, the ecology of rare species, and environmental aesthetics. He is past president of the Ecological Society of America and a board member of the World Wildlife Fund. He is chair of the NRC Board on Environmental Studies and Toxicology and has served on two previous NRC committees, the Committee on the Formation of the National Biological Survey and the Committee on the Applications of Ecological Theory to Environmental Problems, which he chaired. He was elected a member of the National Academy of Sciences in 1989.

Patricia A. Cochran is an Inupiat Eskimo born and raised in Nome, Alaska. Ms. Cochran serves as Executive Director of the Alaska Native Science Commission (ANSC), a cooperative effort of the Alaska Federation of Natives, University of Alaska Anchorage and the National Science Foundation. Ms. Cochran previously served as administrator of the Institute for Circumpolar Health Studies at the University of Alaska Anchorage; Executive Director of the Alaska Community Development Corporation; Local Government Program Director with the University of Alaska Fairbanks; and Director of Employment and Training for the North Pacific Rim Native Corporation.

John W. Duffield is Professor of Economics at the University of Montana where he has taught since 1974. Dr. Duffield's research focus is on nonmarket valuation of wildlife and fishery resources. In recent years he has directed studies of natural resource issues ranging from wolf recovery in Yellowstone N.P. to instream flow levels in Montana trout streams. His work in Alaska includes analysis of lake fisheries in the Tanana Valley and Arctic grayling fisheries in the Arctic-Yukon-Kuskokwim region. In 1992 he participated in the Exxon Valdez litigation on behalf of the Alaska Native class. He is a co-author, with Kevin Ward, of *Natural Resource Damages: Law and Economics* (New York: John Wiley & Sons 1992).

Todd K. Fuller is an Associate Professor of Wildlife Ecology in the Department of Forestry and Wildlife Management at the University of Massachusetts, Amherst. He has conducted research on wolf population ecology and prey relationships in Alberta, and on wolf, white-tailed deer, and hunter interactions in Minnesota. He also has studied the ecology of jackals and African wild dogs in Kenya, foxes in Chile, and black bears, porcupines and fishers in Massachusetts. Dr. Fuller teaches courses in wildlife ecology, conservation, and management, while also guiding graduate students studying a variety of carnivores and herbivores in California, Massachusetts, Minnesota, Vermont, China, Costa Rica, Mongolia, and Papua New Guinea.

Ralph J. Gutierrez is currently a Professor of Wildlife Management at Humboldt State University. He has been studying wildlife since 1968. His primary areas of interest and research are endangered species management, game management, habitat ecology, and hunting ethics. He has conducted both research and consulting in Alaska. In addition to his many research activities and interests, he has served on many government and private conservation committees. He has published many scientific papers in his areas of interest including the book *North American Game Birds and Mammals.*

W. Michael Hanemann is professor of economics at University of California Berkeley. He works on contingent valuation, the most widely used approach in economic assessments of wildlife management issues. He was commissioned by the Alaska Board of Trustees to assess the economic impact of the Exxon Valdez oil spill on Alaskan fisheries.

Frances C. James is a professor in the Department of Biological Science at Florida State University, where she teaches courses on the ecology and systematics of birds and mammals, conservation biology, and data analysis. Her major current research projects involve the analysis of broad scale population trends in birds and the relationship between fire ecology and the endangered fauna of the southeastern pine forests. She is the associate editor of the Annual Review of Ecology and Systematics and served as 1997 president of the American Institute of Biological Sciences. In 1997 she received the Ecological Society of America's "Most Eminent Ecologist" award.

Peter M. Karieva is professor of ecology at the University of Washington, Seattle. Dr. Karieva is a population ecologist whose primary subject is population modeling, but who is also interested field biology. He is currently supervising a Ph.D. student's work on wolf-prey interactions.

Stephen R. Kellert is a professor at the Yale University School of Forestry and Environmental Studies. Much of Dr. Kellert's work has focused on the value of wildlife, including attitudes towards wildlife and human dimensions in wildlife management. He has received awards from the Society for Conservation Biology (Distinguished Individual Achievement, 1990), International Foundation for Environment Conservation (Best Publication of the year, 1987), National Wildlife Federation (Conservationist Of The Year, 1983), and a Fulbright Research Fellowship (Japan, 1985-86). He has served on Agriculture and Wildlife Committees at the National Science Foundation, and is a member of the IUCN Species Survival Commission Specialist Groups. He has published over 100 scientific papers and books including, most recently, "The Value of Life: Biological Diversity and Human Society" (Island Press, 1996), and "Kinship to Mastery: Biophilia in Human Evolution and Development" (Island Press, 1997).

David Klein is Senior Scientist with the Alaska Cooperative Fish and Wildlife Research Unit and Professor in the Department of Biology and Wildlife and Institute of Arctic Biology at the University of Alaska Fairbanks. He worked as a biologist in Alaska with the US Fish and Wildlife Service and the Alaska Department of Fish and Game before joining the faculty of the University of Alaska in 1962. Dr. Klein's research emphases include population ecology, habitat relationships, and foraging dynamics of northern ungulates, primarily caribou, muskoxen, deer, and moose; ecology and adaptations of arctic wildlife; and influences of northern development activities on wildlife. Alaska has been the primary focus of his research activities, although he has also worked in Canada, Greenland, Scandinavia, Siberia, and Africa.

Bruce McLellan has been employed since 1989 as a Wildlife Research Ecologist with the British Columbia Ministry of Forests Research Branch. His work involves the coordination and implementation of wildlife/forestry research on issues that are of provincial importance. Dr. McLellan works on black and grizzly bears in other locations in British Columbia. In addition to his work on bears, he has conducted research on the ecology of caribou, wolves and wolverine in an area primarily managed for timber production for the past 5 years.

Perry D. Olson is a native of Colorado and graduate of Colorado State University. He has over 35 years of professional experience in the field of wildlife management and wildlife administration. He began his professional career with the Colorado Division of Wildlife as a Wildlife Conservation Officer in 1960 and has held a number of positions including Wildlife Biologist, State Big Game Biologist, Regional Director and Executive Director. He retired in 1996 after serving as Executive Director of the Colorado Division of Wildlife for over 7 years. Mr. Olson holds numerous national and international awards for creativity and excellence in wildlife management and administration.

George Yaska is Liaison Officer for the Tanana Chiefs Conference, Inc. (TCC), which is a consortium of 43 tribes within the Interior of Alaska. Before this September, he was Director of Wildlife and Parks at TCC. The area TCC encompasses is approximately equal to 39% of Alaska.

He is trained and experienced in traditional knowledge. He has spoken with hundreds of Alaska Natives and has also been trained by his family and community.

NRC STAFF

Janet Joy *(Study Director)* has been a Program Officer at the National Research Council since 1994. She received a BS in behavior and biology from the University of Michigan, a Ph.D. in zoology from the University of Toronto in 1983, and did postdoctoral work at the University of Texas and Northwestern University. She was a Senior Staff Fellow at the National Institute of Mental Health at NIH from 1989-1994. Her research interests include environmental biology, behavior, and neuroscience. At the NRC, she has served as study director for projects on biodiversity management and intellectual property issues.

Jeff Peck *(Project Assistant)* worked on this study from January 1996 to January 1997. In his four years at the Board on Biology, he assisted with numerous studies including The Value of Biodiversity, Intellectual Property Rights and Research Tools in Molecular Biology, the Scientific Basis for the Preservation of the Mariana Crow, and the Convocation on Scientific Conduct. He has a BA in journalism, and a Masters in international journalism, both from Baylor University. He has written several freelance articles including political and international news features.

Allison Sondak *(Project Assistant)* is a project assistant at the Board on Biology. She has worked on fisheries and environmental policy in the US Senate and on park management at the National Park Service. Her interests include natural resource policy and coastal zone management. She earned a BA in Environmental Policy at Duke University, and has studied marine science at the University of Copenhagen, Denmark.

APPENDIX

C

Wolves and Caribou in GMU 20: Example of Assessing Predator-Prey Dynamics by Testing the Fit of Different Models to Available Data

Data on wolves and caribou in Alaska are among the most extensive for any large mammal predator-prey interactions. Given this, one might think it easy to determine the impact of wolves on their prey populations. More specifically, game managers should be able to use such data to determine:

• The extent to which wolves keep prey populations below their carrying capacity.
• Whether predator control can increase prey populations without extensive killing of predators that must be done year after year.

Well-designed experimental manipulations offer the clearest way of measuring the impact of wolves on their prey, but they are not always feasible. As an alternative to experimentation, one can propose different models or hypotheses to see how well they "fit" or "explain" the data. To assess how much inference is possible from such model fitting, the committee asked ADFG biologists for their most complete long-term records of wolf and caribou numbers, including data collected during a period of wolf control. Patrick Valkenburg provided population data for GMU20A, including unpublished data. Those data span the years 1970 to 1995 and consist of population counts of caribou and wolves, number of wolves killed, number of caribou killed by hunters, calf:cow ratios for different months, caribou births, and information about weather such as summer temperatures, rain, and snow depths.

One way of asking whether wolves limit caribou populations is to search for correlations between the number of wolves and calf survival (represented by the number of calves in the fall divided by the number born in the spring). Figure C.1 shows that calf survival declines as the number of wolves increases. One might be tempted to conclude from such a correlation that in this particular wolf-caribou system wolves control the number of caribou. However, one can find other correlations that support very different conclusions. For example, figure C.1 also shows that the amount of summer rainfall is positively correlated with calf survival. Thus these simple correlations, in which one variable is plotted against another variable, do not tell us which is more important to calf survival: the impact of wolves or the impact of weather. Indeed, to a large extent, the debate about the role of wolves in caribou population dynamics arises from the ambiguity inherent in using simple correlations to determine what factors limit prey populations.

One way to weigh the relative importance of different potential limiting factors is to use all the wolf-caribou data in a single analysis, and to ask which possible hypotheses best explain the total pattern of the data, as opposed to a subset of the data used in a single correlation. For example, one can write a population model in which the number of calves killed per unit time is linearly related to the number of calves available and the number of wolves present. Using this quantitative model of wolf predation, one can then ask how much of the variation in caribou numbers can be explained simply by wolf predation. The results of such an analysis are shown in figure C.2. The parameters that describe the graph relating calf survival to wolf numbers were selected to provide the best fit to the data. (In this case, a curve is fitted to the data, and "best fit" is defined as the equation for the curve with the smallest sum of the squared differences between observed data and the predicted data on caribou numbers and calf:cow ratio.)

The model, indicated by the solid line in figure C.2, clearly identifies the main trends in the data shown in the open circle. The simple assumption that wolf predation increases linearly with number of caribou is sufficient to account for the major trends in caribou data, without taking any other factors into account. One can then test other factors in the same way. Indeed, none of the weather data could be correlated with the pattern of up and down trends in caribou numbers seen between 1970 and 1995. For an example, see the analysis of snow depth in figure C.3. Although, the model depicted in figures C.2 and C.3 is quite simplistic, it nonetheless suggests a direction of quantitative research that might help wildlife manages learn more from their data than they can by looking for simple correlations between 2 factors at a time. The procedure is straightforward:

1. Propose a specific model for caribou population dynamics.
2. Estimate rates in that model either directly from field data, or by fitting parameters such that the observable variables (such as predator and prey numbers, calf survival, and weather data) are as well-described as possible (by a formal criterion such as least-squares, or maximum likelihood).

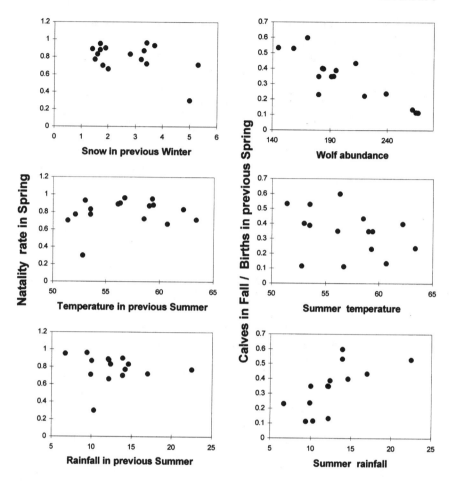

FIGURE C.1   Correlations between key caribou demographics (natality and calf surviv-al as measured by calves counted in the fall divided by spring births) and several possible causal factors (such as snow depth, number of wolves, summer temperatures and rainfall, previous summer rainfall). All plots are derived from the same data, albeit with different possible independent variables (causal factors).

3. Evaluate competing hypotheses by comparing their ability to describe trends in caribou numbers.

The approach described above differs from the simple correlations depicted in figure C.1 because the changes in population counts over time are analyzed in relation to several factors that are known to be acting concurrently. These time series are ordered and structured. When the analysis is reduced to a single index

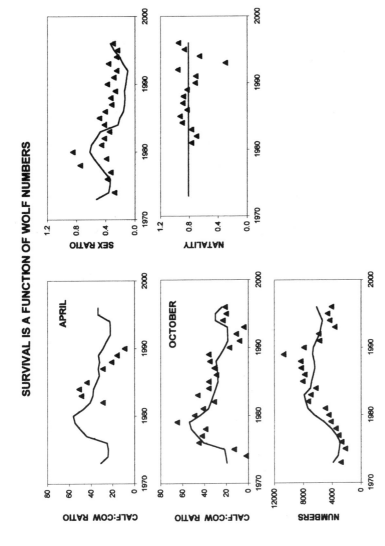

FIGURE C.2   The results of a simple model of caribou population trends assuming constant rates of natality, and survival that varies only as a result of wolf predation.   Other data used in the model, such as hunting mortality for caribou, were provided by ADFG biologists.   The solid line indicates the best fit wolf predation model, that is, the model that is most consistent with the trends observed in the data.

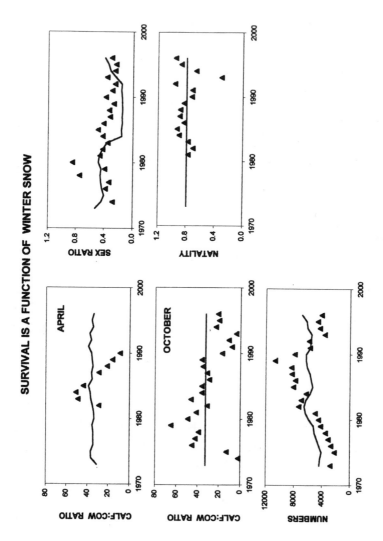

FIGURE C.3   The results of a best fit caribou model in which the only parameter change through time is caribou survival which is assumed to show a straight line correlation with snow depth. Calf survival is assumed to be constant in this model regardless of the number of wolves. The model is not consistent with the patterns seen in the data such as the widely varying calf:cow ratios in October, although it is consistent with varying calf:cow ratios in April.

of survival in a correlation test as in figure C.1 this order and structure is lost, because the information about actual year or sequences in which the caribou data were collected is ignored.

It is important to note that this model fitting can still lead to ambiguous answers. In such cases, one can choose management actions that help to discriminate between competing hypotheses. The strategy of adaptive management is to choose actions that increase what is known about the system being managed. This is also best done if there is an actual population model, and not just simple correlations. A key point is that only by writing an explicit model for population change can one adequately measure the impact of predators on prey and interpret the results of different management actions. Otherwise, one is too vulnerable to the illusion that the simple correlations observed between predators and their prey are true indicators of predator-prey dynamics.